SALT INTRUSION, TIDES AND MIXING IN MULTI-CHANNEL ESTUARIES

Cover design: Peter Stroo.

Cover photos: Satellite images of the Scheldt estuary and the Mekong Delta branches (left to right), source: Google Earth and ESA.

Salt Intrusion, Tides and Mixing in Multi-channel Estuaries

DISSERTATION

Submitted in fulfilment of the requirements of

the Board for Doctorates of Delft University of Technology

and of the Academic Board of the UNESCO-IHE Institute for Water Education

for the Degree of DOCTOR

to be defended in public

on Friday, 22 February 2008 at 10.00 hours in Delft, The Netherlands

by

NGUYEN Anh Duc

Master of Science in Hydraulic Engineering (with distinction),

UNESCO-IHE Institute for Water Education, Delft, The Netherlands

born in Hanoi, Vietnam

This dissertation has been approved by the promotor:

Prof. dr. ir. H.H.G. Savenije TU Delft/UNESCO-IHE, The Netherlands

Members of the Awarding Committee:

Chairman	Rector Magnificus Delft University of Technology
Vice-chairman	Director UNESCO-IHE Institute for Water Education
Prof. dr. ir. H.H.G. Savenije	TU Delft/UNESCO-IHE, promotor
Prof. dr. ir. N.C. van de Giesen	Delft University of Technology
Prof. dr. ir. J.A. Roelvink	UNESCO-IHE Institute for Water Education
Prof. dr. ir. G.S. Stelling	Delft University of Technology
Prof. dr. ir. M.J.F. Stive	Delft University of Technology
Dr. ir. Z.B. Wang	Delft University of Technology

Dr. T.D. Thang, Southern Institute for Water Resources Research in Vietnam, and Ir. M. van der Wegen, UNESCO-IHE Institute for Water Education, have provided substantial support in the preparation of this dissertation.

Taylor & Francis is an imprint of the Taylor & Francis Group, an informa business

Published by:
Taylor & Francis/Balkema
PO Box 447, 2300 AK Leiden, The Netherlands
e-mail: Pub.NL@tandf.co.uk
www.balkema.nl, www.taylorandfrancis.co.uk, www.crcpress.com

ISBN 978-0-415-47122-0 (Taylor & Francis Group)

To my parents

PREFACE

I got the first academic impressions about salt intrusion in July 2002, when some friends of mine studying in Coastal Engineering (UNESCO-IHE) told me about the interesting subject "Density currents and salt intrusion in estuaries", which was taught by Ir. Mick van der Wegen and Prof. Hubert Savenije. Unfortunately, I did not have a chance to attend the class since I was following another one at the same time, and I thought I should take it in the near future.

Shortly after, when carrying out my M.Sc. study in flood management of the lower Dong Nai – Sai Gon river basin in Vietnam, I realized that apart from floods and water quality, salt intrusion is a major problem in the basin. The problem appeared even more severe in the neighbouring system, the Mekong Delta, which is the most important agricultural and aquacultural area of South Vietnam. I was inspired by the idea that I would carry out research on the salt intrusion in the lower Dong Nai – Sai Gon river basin and the Mekong Delta. I was fortunate to have a number of people to encourage and to support me following this idea. My M.Sc. study supervisors, Ing. Klaas-jan Douben and Prof. Bela Petry, and my employer in Vietnam, Dr. Tang Duc Thang introduced me to Prof. Savenije, who is an expert in salt intrusion and estuarine hydraulics. After their introduction, I went to see Prof. Savenije in person and to express my idea to do some research on the subject of salt intrusion. With his enthusiasm and his inspiring vision about solving the salt intrusion problem and analyzing tidal hydraulics, we plotted the plan for the research proposal and for getting the research fund.

After the funding issue was settled, I started my research in February 2004. It took ten months to finalize the research proposal, during which I learned much about not only salt intrusion, but also mixing and tidal hydraulics. The Mekong Delta and the Scheldt estuary became the main study areas of the research.

In the beginning of 2005, I went to the Center of Water Research, University of Western Australia, Perth. The two-month period working there was a great success. I and Prof. Savenije worked very closely together in defining the direction of the research. The proposed salinity measurement campaigns for the Mekong Delta using the moving boat method were shaped up. The first ideas for a paper on salt intrusion were formulated. I enjoyed the working environment there and had a number of occasions to meet and discuss things with Prof. Murugesu Sivapalan, Prof. Jörg Imberger and their Ph.D. students.

The measurement campaigns in the Mekong Delta in 2005 and 2006 happened relatively smoothly and successfully, although it was a real challenge to use a small speedboat in such a big estuary with strong tides and wind. We managed to carry out measurements in four branches of the Mekong with the assistance of several local staff and my brother-in-law. The campaigns taught me a valuable lesson about how complicated the system is.

The last two and half years of my research went really fast. Most of the time I focused on running models, writing papers, and shaping up my thesis. I was helped by a number of people to speed up my research: Ir. Mick van der Wegen, Dr. Tang Duc Thang, Dr. Zheng Bing Wang, Prof. Dano Roelvink, and Ir. Adri Verwey; not to mention Prof. Savenije for his guidance and support on the regular basis. I had chances to attend several national and international conferences in The Hague, Vienna, Varna, Alexandria and Phoenix and to meet people there to discuss research interests. I also had a number of occasions to deal with several tough but constructive reviews to improve my journal papers and my research.

Overall, I am satisfied with the results of my research and I am pleased to have the dissertation finished in time. However, I can see there is a need to further study salt intrusion, mixing and tidal hydraulics in estuaries, since there are so many beautiful things waiting for us to explore.

Nguyen Anh Duc,

Delft, February 2008.

SUMMARY

An estuary is a source of food and a transport link between a river and a sea. The estuary therefore has characteristics of both the river and the sea, being a unique environment influenced by tidal movements of the sea and freshwater flow of the river. Some estuaries are classified as multi-channel estuaries. A multi-channel estuary has at least two branches. Each branch functions as a single estuary, but highly interacts with its neighboring branch. In this study, we concentrate on two multi-channel estuaries. The first one is the Mekong Delta in Vietnam, which is a delta estuary consisting of eight branches. The second estuary is the Scheldt in the Netherlands, which has a funnel shape with a distinct ebb-flood channel system.

Tides transport salt-water in and out of an estuary and mix it with fresh river water. The mixing process in an estuary is complex and much depends on the estuary characteristics. Tide-driven and density-driven mixing are the most important mechanisms. The river flow drives the density-driven circulation, accompanied by vertical salinity stratification. Tidal pumping appears to be an important tide-driven mixing mechanism. Tidal pumping caused by large-scale ebb-flood channel residual circulation is important in estuaries with a distinct ebb-flood channel system, however this mechanism has not been researched much to date.

Tide-driven and density-driven mixing cause salinity to intrude further inland. The salt intrusion can reach a large distance from the coastline and affect water-use activities in the estuary. Although the salinity intrusion has been well studied in single-branch estuaries, for a multi-channel estuary, to date a predictive steady-state has not been developed.

The objectives of this study address a number of knowledge gaps: (i) to investigate and to develop a predictive steady state salt intrusion model for a multi-channel estuary; (ii) to develop a new approach for estimating the distribution of freshwater discharge over the branches of a multi-channel estuary; (iii) to analyse characteristics of tidal waves in multi channel estuaries; and (iv) to develop a theory analysing effects of tidal pumping caused by residual ebb-flood circulation on salinity distribution in multi-channel estuaries and to propose a new analytical equation to quantify the 1-D effective salt dispersion coefficient for the tidal pumping mechanism.

Firstly, on the basis of salinity measurements carried by moving boat method during the dry season of 2005 and 2006, salinity intrusion of the Mekong Delta branches has been computed. An analytical model, based on the theory for the computation of salt intrusion in single branch alluvial estuaries, has been developed for multi-channel estuaries. This model has been successfully applied and tested in the branches of the Mekong Delta. The model has been then validated with data of the dry seasons in 1998 and 1999. On the basis of these results, a predictive steady-state model has been developed. The overall results of the salinity computation are good, indicating that this model produces satisfactory results for a complex system such as the Mekong Delta.

Secondly, a new approach for determining freshwater discharge distribution over the branches of the Mekong Delta has been presented. The freshwater discharge is an important parameter for modelling salt intrusion. The determination of the fresh water discharge in estuaries is complicated, as it requires detailed measurements during a full tidal cycle, whereby the accuracy is low. In the past, only numerical models could provide this information. On the basis of the newly developed predictive salt intrusion model, a new approach has been developed to determine the freshwater discharge distribution by means of salinity measurements. The new approach has been compared with the most recent 1-D hydrodynamic model of the Mekong river system. The comparison shows that the new approach provides a good picture of the discharge distribution and can be a powerful tool to analyze the water resources in tidal regions.

Thirdly, tidal characteristics of the Mekong Delta and the Scheldt have been explored. The three main factors, i.e. tidal wave celerity, phase lag and tidal range variability, have been investigated. The tidal wave characteristics have been analysed in the two main sub-systems of the Mekong Delta: the Tien and the Hau, mainly by analytical equations describing tidal wave characteristics and the 1-D hydrodynamic model. The agreement between these model approaches and observations is reasonable, especially for the less complex Hau system. It appears that the Mekong Delta branches have a small estuary shape number, therefore have a large phase lag and a damped wave. Hence, the Mekong Delta branches are mostly riverine in character. For the case of the Scheldt, the agreement between the two model approaches and observations is very good. The Scheldt estuary is a marine estuary with a large estuary shape number, a smaller phase lag and an amplified wave.

Fourthly, an analytical equation has been derived to determine the longitudinal dispersion coefficient in an estuary, where tidal pumping due to ebb-flood channel residual circulation is the dominant mixing process. The newly developed equation takes into account two important factors of the residual circulation, i.e. the tidal pumping efficiency and the ebb-flood loop length. Mixing caused by gravitational effects associated with marked stratification is not addressed by the equation; this can be obtained by the method of Harleman and Thatcher. A 3D hydrodynamic model in DELFT3D has been employed to generate a solution of the circulation pattern for the Western Scheldt estuary in the Netherlands. This solution has been subsequently decomposed to isolate the influence of different hydrodynamic processes on mixing. The analytical equation then has been compared with the results from the 3D model that represented tidal pumping, and reasonable agreement has been found. Comparison has also been made between the 3D model, the analytical equation, and a steady-state salt intrusion model. Finally, observed data from the Western Scheldt estuary and the Pungue have been employed to validate the calculated dispersion values. The good performance of the newly developed equation in comparison with the existing models as well as with observations indicates that the equation is indeed applicable in practice. The new equation provides an opportunity to evaluate the large-scale mixing mechanism caused by the ebb-flood channel residual circulation, which is not feasible to analyse from field observations.

The study has successfully developed a predictive analytical approach for analysing salt intrusion and mixing in multi-channel estuaries. The new approach compares well with hydrodynamic models and observations, indicating its applicability in practice. Most importantly, the study has developed a new equation to quantify tidal pumping due to ebb-flood channel residual circulation in terms of a longitudinal salt dispersion. Recommendations have been made to further apply the predictive approach to other multi-channel estuaries. The unique tidal wave characteristics in multi-channel estuaries should be further investigated.

CONTENTS

Chapter 1

INTRODUCTION

1.1 INTRODUCTION

Estuaries have always been important to mankind. An estuary is both a source of food and a transport link between a river and a sea. Almost every large estuary in the world is the site of a major city, especially for port and transport development. In estuaries, freshwater collected over vast regions of the land pours into a sea or an ocean, which sends salt water upstream far beyond the river mouth. Vigorous mixing between the two fluids creates a unique environment, with large potential for life forms able to handle the associated large variability in environmental conditions.

An estuary has characteristics of both a river and a sea. The sea and the river exchange their water, substances and sediments. The estuary is, therefore, a unique environment that is mainly influenced by tidal movements of the sea and freshwater discharge of the river.

Tides propagating in an estuary are generally a mixture of progressive waves and standing waves. Tides carry salinity and other substances (e.g. nutrient, sediment, etc.) in and out of the estuary as well as mix them over the entire estuary. Saline water can intrude inland due to the difference in density between the freshwater and the seawater and due to tidal movement. The salinity intrusion itself can reach a great distance from the coastline, especially when the river flow is small. It affects every water-use activity in the estuary, e.g. domestic, agricultural, industrial and other uses; therefore it may damage the interests of people of very large areas in the estuary. Thus, prediction of salinity intrusion in estuaries has received a lot of attention by researchers.

Mixing mechanisms in estuaries are quite complex and they depend much on the estuarine characteristics. In a single-branch stratified estuary, mixing by gravitational circulation can play a dominant role, but in another single-branch well-mixed estuary, tidal pumping can be the most important mixing mechanism. In multi-channel estuaries, the same thing happens, but in a more complex way due to the complicated topography. Hence it is a challenge to analyse mixing mechanisms and salt intrusion in multi-channel estuaries.

1.2 MAIN STUDY AREAS

1.2.1 Mekong Delta

The Mekong river is the longest river in South Asia and the twelfth longest in the world. The Mekong river basin covers a catchment of approximately 795,000 km^2. As the economy of this region has developed at a high growth rate in recent years, the Mekong river basin faces complicated problems not only in water quantity and quality, but also in ways of usage. The issue of sustainable development and management of the basin to meet not only the economic needs but also the social, cultural and environmental needs has become one of the top priorities (Cogels, 2005; and Trinh and Nguyen, 2005).

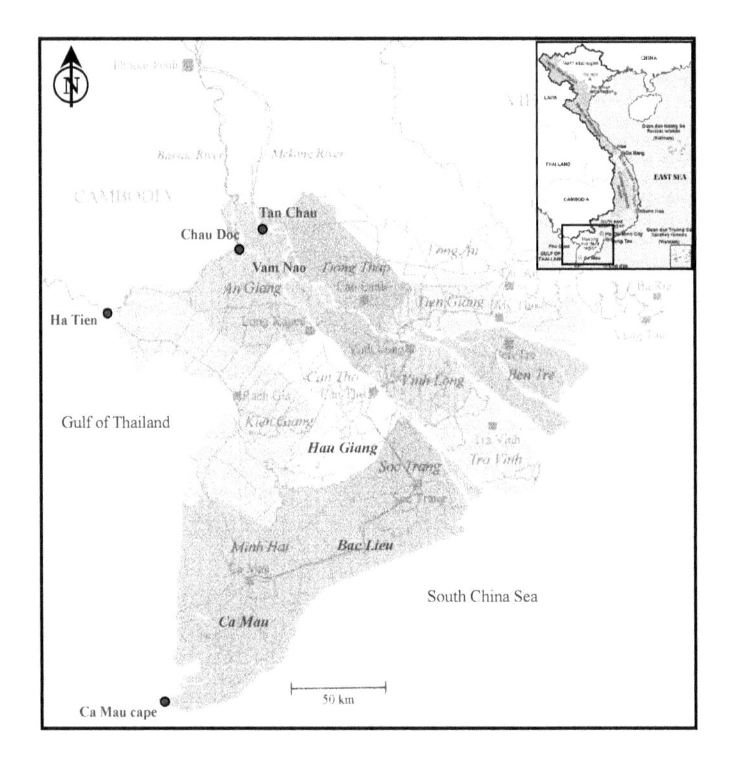

Figure 1.1 Map of the Mekong Delta

Originating from China, the Mekong river and its tributaries drain regions in six countries: China, Burma, Laos, Thailand, Cambodia and Vietnam. The southern part of the Mekong river basin, downstream of Burma - Laos - Thailand border, is defined as the lower Mekong basin. The Mekong river when it enters Vietnam splits into two branches, the Bassac (known as the Hau river in Vietnam) and the Mekong (known as the Tien river in Vietnam). These two branches form the Mekong Delta (see Fig. 1.1). The Hau river is the most southern branch of the river system. When the Hau approaches the sea, it splits into two sub-branches: Tran De and Dinh An. The Tien river is the northern branch of the river system, which separates into two sub-branches at Vinh Long: Co Chien and My Tho. At a distance of 30 km from the

South China Sea, the Co Chien river again splits into two smaller branches, Co Chien and Cung Hau. In the downstream part, the My Tho river separates into four branches: Tieu, Dai, Ba Lai and Ham Luong (see details in Fig. 3.2, Chapter 3).

Most of the Mekong Delta is situated within the border of Vietnam, covering 13 provinces: Long An, Tien Giang, Ben Tre, Dong Thap, Vinh Long, Tra Vinh, An Giang, Can Tho, Hau Giang, Soc Trang, Bac Lieu, Ca Mau and Kien Giang (see Fig. 1.1). The Delta covers 39,000 km^2, harbouring some 15 million people. The Delta's rich resources are of vital importance to Vietnam, they account for some 40% of agricultural production in the country, including 50% of the rice production (Trinh and Nguyen, 2005). Rice and fishery products contribute significantly to export earnings and account for about 27% of the Gross Domestic Product.

The hydraulic and hydrological regimes in the Mekong Delta depend on upstream discharge, local rainfall and tidal movement of the South China Sea and the Western Sea (Gulf of Thailand). However, during the dry season when the salinity intrusion problems prevail, local rainfall plays a very minor role (Le, 2006). Thus, the delta is mostly affected by both the river flows and the tidal movements.

The discharge distribution between two branches of the river: the Mekong (i.e. Tien) and Bassac (i.e. Hau) at Tan Chau and Chau Doc is unequal. An annual ratio for Tan Chau/Chau Doc could be roughly estimated at 83%/17%. The ratio is lower in the flood season (80%/20%) and higher in the dry season (84-86%/16-14%) (Le, 2006). The Vam Nao channel is a connecting river and it supplies water for the Hau river. The role of the Vam Nao channel is to balance the water flow between the Hau and Tien river and it plays an important role in the hydraulic regime of the entire Mekong Delta.

The tidal movement at the various estuarine mouths has an important impact on drainage and salinity intrusion. The coastal region from Vung Tau to the Ca Mau cape is affected by tide of the South China Sea, whilst the coastal region from the Ca Mau cape to Ha Tien is affected by the tide of the Western Sea (Gulf of Thailand). Tides in the South China Sea have a mixed diurnal and semi-diurnal character with a period of 12.25 hours. There are generally two troughs and two peaks during a day, but their relative height varies over a fortnight. When the first trough decreases from day to day, the other trough increases, and vice versa. The tidal range is relatively high, about 2.5 - 3.5 m depending on the location (Le, 2006; and Tang, 2002). In general, the tide in the South China Sea varies daily, seasonally as well as yearly. Beside these periods, multi-year variations also take place, but the differences are small and can be ignored.

Exchange between the Mekong river and the sea causes a number of problems, especially in salinity and sedimentation. Before 1980, every year in the dry season, agricultural areas in the Mekong Delta were affected by salinity, amounting to 1.7 – 2.1 million ha out of 3.5 million ha in total (MRC, 2004). In the 1980's and 1990's, a number of projects for salinity control were implemented. They can be listed as: Go Cong (to protect 26,000 ha of agricultural areas from salinity), Tiep Nhat (11,000 ha), South Mang Thit (293,000 ha), Quan Lo – Phung Hiep (290,000 ha), Western Sea coast (50,700 ha) etc. Nowadays, salinity affects only 800,000 – 900,000 ha every year. However, fresh water intakes in the Mekong branches are usually affected by salt-water intrusion. Every year, these intakes have to be closed for quite long periods (from some weeks to one or two months) to prevent salinity intrusion (Nguyen and Nguyen, 1999).

On the other hand, with the new strategies for economic development in the Mekong Delta in Viet Nam and with the principles for diversity of economic activities, areas reserved for aquaculture and shrimps in saline and brackish water are increasing not only in the non-protected areas but also in the protected areas. The demand area for aquaculture and shrimps is more than 300,000 ha (Tang, 2002). The need for coordinating and controlling the saline water for aqua-cultural purposes and fresh water for agricultural purposes is very urgent.

1.2.2 Scheldt estuary

The Scheldt river has its origin in France but its basin is mainly located in Belgium and the Netherlands, draining about 21,580 km^2 of land in one of the most densely populated (10.5 million people) and highly industrialized regions of Europe. The total length of the Scheldt river, including both its estuary and upper river, is 355 km. The major tributaries of the estuary are the Rupel, Durme (both with tidal-influence) and the Dender (closed-off and hence non-tidal) (see Fig. 1.2).

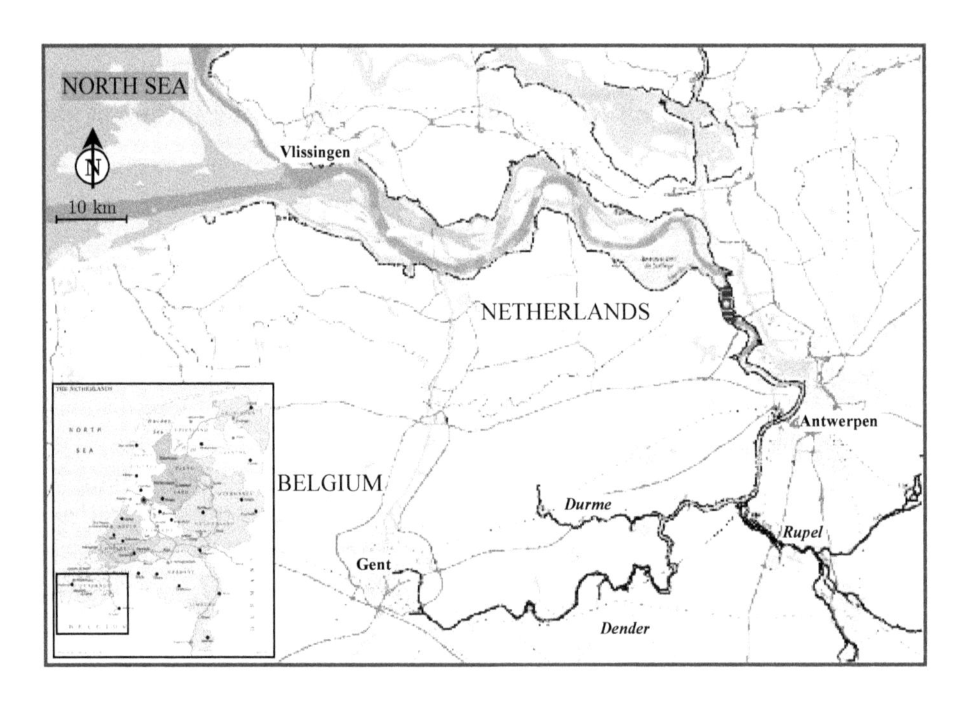

Figure 1.2 Map of the Scheldt estuary

The Western Scheldt is the downstream part of the Scheldt estuary, located in the southwestern part of the Netherlands. The Scheldt estuary has a length of 200 km and stretches up to Gent in Belgium, where the estuary is closed by sluices (van Eck, 1999). It has a pronounced funnel-shape geometry and covers an area of about 370 km^2. Its cross-sectional area and width decreases exponentially from the estuary mouth at Vlissingen to the estuary head near Gent. The estuary can be characterized as a multi-channel estuary with a regular system of ebb and flood channels (Van Veen *et al.*, 2005).

The main functions of the Scheldt estuary are navigation, ecology, recreation and fishery. The Scheldt estuary region is both important agricultural and industrial area. The estuary is used as a major drain for industrial and domestic wastes (Baeyens *et al.*, 1998). A substantial part of these is still not treated in a wastewater treatment plant, resulting in poor water quality in some parts of the estuary (Verlaan, 1998).

The hydraulic and hydrological regimes in the Scheldt estuary depend on upstream discharge, local rainfall and tidal movement of the North Sea. Tides in the Scheldt can be regarded as semi-diurnal meso-tide to macro-tide. The spring tidal range increases from about 3.8 m at Vlissingen (at the mouth) to 5 m at Antwerpen (78 km upstream) and decreases till Gent. The annually average river discharge is about 120 m^3/s, which is generally very small in comparison to the tidal flows. However, variations from year to year can be large, ranging between 50 and 200 m^3/s (Kruijper *et al.*, 2004).

Due to the increasing human activities, demands on improving the conditions of the Scheldt estuary have been raised (van Maldegem, 1993; Arends and Winterwerp, 2001). It is worthwhile to investigate the mixing mechanism for moving pollutants and transporting sediments and salinity in the Scheldt estuary in order to have a better understanding of the system.

1.3 RESEARCH CHALLENGES AND OBJECTIVES OF THE STUDY

Firstly, it is a major challenge to predict the salt intrusion regime in multi-channel estuaries, such as the Mekong Delta in Vietnam, in order to adequately manage salinity. For a 1-D predictive model, one needs adequate hydraulic parameters and a predictive theory on mixing processes as a function of the river discharge and tidal regime. In particular, the dispersion coefficient should be predictable. To date, a model has not yet been developed to cope with salinity intrusion in multi-channel estuaries such as the Mekong Delta. Moreover, existing tools that may be used to compute salinity intrusion are not predictive since they require calibration of the dispersion coefficients.

Secondly, it is obvious that tides and fresh water discharge affect the salinity regime in estuaries. Reversely, the salinity regime reflects how the freshwater discharge distributes over the entire system. In the Mekong Delta, it is a challenging task to estimate how the discharge freshwater distributes over the branches of the multi-channel estuary and hence to see if it is possible to use salinity data to make such an estimation.

Thirdly, tides propagating in an estuary are a mixture of progressive and standing waves, which experience feedbacks from topography and friction. It is interesting to analyse the tidal wave characteristics in multi-channel estuaries and to see how the multi-channel topography affects the tidal movements.

Finally, the interaction between tides, freshwater and topography causes mixing in alluvial estuaries. Especially for multi-channel estuaries having characteristics of a distinct ebb-flood channel system, such as the Scheldt and the Pungue in Mozambique, large-scale residual ebb-flood channel circulation is an important mechanism for moving pollutants and transporting salinity upstream against a mean discharge of fresh water, for which no adequate theory exists as yet.

In view of these challenges, the objectives of this study are:

(i) To investigate and to develop a predictive steady state salt intrusion model for a multi-channel estuary.

(ii) To develop a new method for estimating the distribution of freshwater discharge over the branches of a multi-channel estuary.

(iii) To analyse characteristics of tidal waves in multi channel estuaries.

(iv) To develop a theory analysing effects of tidal pumping caused by residual ebb-flood channel circulation on salinity distribution in multi-channel estuaries and to propose a new analytical equation to quantify the 1-D effective salt dispersion coefficient for the tidal pumping mechanism.

1.4 RESEARCH APPROACH AND OUTLINE OF THE THESIS

This study approaches its objectives via several steps. Firstly, a literature study is carried out. It provides an overview on estuaries, tides, mixing and salinity intrusion. The existing information reveals that there are not many studies done for multi-channel estuaries. Secondly, through an extensive quest for the Mekong Delta data, it appears that the observations on salinity are only available for few fixed stations spreading over the delta. Taking into account the complexity of the system, it is not possible to develop a predictive steady state salt intrusion model for the Mekong delta on the basis of the existing data. Therefore field campaigns have to be carried out on salinity measurements, preferably using the moving boat methods to obtain simultaneous longitudinal salinity distributions. Thirdly, similar to the salinity data, observations on discharge and water level are not sufficient due to a very small number of stations. With the respect to the discharge, there are only five upstream discharge stations, however, there is no station in the downstream – saline parts of the delta. Due to the lack of necessary observations, it is very difficult to get an estimation on how the discharge distributes over the branches of the delta. This estimation can be obtained on the basis of results from a predictive analytical salinity model, and it is essential to check the results with a hydrodynamic model in order to indicate the applicability of the analytical solutions. Fourthly, in order to obtain a good view on tidal wave propagation in multi-channel estuaries, one needs detailed spatial and temporal observations. It is mentioned earlier that sufficient data is not available in the Mekong delta. The Scheldt estuary has a better data set due to the dense station network; however, it is still not good enough to explore the longitudinally tidal characteristics of the system. Thus, use has been made of well-calibrated 1-D hydrodynamic models and analytical equations for analysing the two systems. Finally, for ebb-flood channel estuaries such as the Scheldt estuary, using observations to investigate large-scale mixing mechanisms such as tidal pumping is not possible. The reason is that it requires a huge set of spatial and temporal observations over the entire system and in reality it is not feasible to get such a dataset. Therefore, a good 3-D hydrodynamic model has been used as a "virtual laboratory" to generate sufficient data for investigating mixing mechanisms.

The thesis can broadly be divided into three parts: Introduction and Overview (chapters 1 and 2); Research results (Chapters 3, 4, 5 and 6); and Discussions and Conclusions (Chapter 7).

Chapter 1 presents a brief introduction on estuaries, tides, mixing and salinity intrusion in alluvial estuaries. It also gives general information of the main studied multi-channel estuaries, i.e. the Mekong Delta in Vietnam and the Scheldt estuary in the Netherlands. It raises the research challenges and defines the objectives of the study. Chapter 2 provides an overview on tides, mixing and the salinity intrusion phenomenon. The overview is essential before proceeding into further studies. Firstly, the classification of estuaries is summarized to get a good view on general characteristics of estuaries and to see how the study areas are. Secondly, mixing mechanisms, which drive saltwater intrusion further inland, are listed. Tools and mathematical models to investigate and simulate mixing and salinity intrusion mechanisms are presented as well as briefly assessed. Particularly, attention is paid to steady state salt intrusion models and mixing caused by residual circulations. Finally, tidal wave characteristics are introduced together with analytical tools to investigate the characteristics.

Chapter 3 shows the development of the new 1-D predictive analytical model for multi-channel estuaries. The model is applied to the case of the Mekong Delta for 1998, 1999 and 2005. Chapter 4 presents the new method to estimate the freshwater discharge distribution in multi-channel estuaries, based on the newly developed 1-D predictive analytical model for multi-channel estuaries. The 2005 and 2006 data of the Mekong Delta are used to validate the new method. Results of the new method are compared to the results of a hydrodynamic model (MIKE11) of the Mekong Delta as well as results of several previous studies. Chapter 5 investigates tidal wave characteristics in multi-channel estuaries. Making use of observations, a set of analytical equations and hydrodynamic models (i.e. MIKE11 and SOBEK-RE), the characteristics of tidal waves in the Mekong Delta and the Scheldt estuary are explored and several interesting conclusions are drawn. Chapter 6 presents the development of the new theory on the mixing phenomenon in multi-channel estuaries caused by large-scale residual ebb-flood channel circulation. One conceptual model and one new analytical equation to determine the effective dispersion coefficient are proposed. The results of the newly developed equation are confronted with the case of the Scheldt estuary using the mathematical model DELFT3D and observations.

Chapter 7 presents the recommendations and conclusions of the study. Recommendations are made for further considerations on application of the predictive salt intrusion model, the method determining the freshwater discharge distribution and the analytical equation computing dispersion coefficient to other systems in the world. Conclusions are given to remark that the study has fulfilled its objectives by developing several new methods for estimating salinity intrusion, freshwater discharge and mixing dispersion in multi-channel estuaries.

Chapter 2

SALT INTRUSION, MIXING AND TIDES IN ESTUARIES

2.1 INTRODUCTION

In this chapter, we present an overview on classification of estuaries, mixing, salinity intrusion mechanisms and tidal wave characteristics in estuaries. The information provided is essential for studying and analysing mixing, salt intrusion and tides in alluvial estuaries.

Firstly, in Section 2.2, the classification schemes of estuaries are summarized to obtain a perspective on general characteristics of estuaries and to see how the study areas fit in these schemes. Secondly, the different mixing mechanisms that drive salt-water intrusion are listed in Section 2.3. Tools and mathematical models to investigate and simulate mixing and salinity intrusion mechanisms are presented as well as briefly assessed. Particular attention is paid to steady state salt intrusion models and mixing caused by residual circulation. Subsequently, in Section 2.5, tidal wave characteristics are introduced together with analytical tools to investigate these characteristics. Finally, conclusions for the literature overview are made in Section 2.6.

2.2 CLASSIFICATION OF ESTUARIES

In this section, an introduction on classification of estuaries is presented. Estuaries have been long studied and classified based on their oceanographic, geomorphology, estuarine stratification or tidal characteristics. Although a number of excellent reviews have been well documented (Fischer *et al.*, 1979; Dyer, 1997; or Savenije, 2005), it is worthwhile to mention some main classification schemes of estuaries herein, since it is important to understand general characteristics of estuaries in order to proceed into further studies and analyses for mixing mechanisms, tidal wave characteristics and salinity intrusion of the main study areas, i.e. the Mekong Delta in Vietnam and the Scheldt estuary in the Netherlands.

Estuaries have characteristics of both a river and a sea; and certainly an estuary is a transition zone between the river and the sea. Savenije (2005, p. 3) showed a clear linkage between an estuary, a river and a sea (see Table 2.1).

Table 2.1 Characteristics of a sea, an estuary and a river (based on Savenije, 2005)

	Sea	Estuary	River
Shape	Basin	Funnel	Prismatic
Main hydraulic function	Storage	Storage and transport	Transport of water and sediments
Flow direction	No dominant direction	Dual direction	Single downstream direction
Bottom slope	No slope	Very small or virtually no slope	Downward slope
Salinity	Saline	Brackish	Fresh
Wave type	Standing	Mixed	Progressive
Ecosystem	Nutrient poor, marine	High biomass productivity, high biodiversity	Nutrient rich, riverine

Table 2.2 Oceanographic classification of estuaries (after Pritchard, 1967)

Estuarine type (*)	Dominant mixing force	Mixing energy	Width/depth ratio	Salinity gradient	Mixing index (**)
A	River flow	Low	Low	Longitudinal vertical	>=1
B	River flow, tide	Moderate	Moderate	Longitudinal vertical, Lateral	< 1/10
C	Tide, wind	High	High	Longitudinal Lateral	< 1/20
D	Tide, wind	Very high	Very high	Longitudinal	-

(Continued)

Estuarine type (*)	Turbidity	Bottom stability	Biological productivity	Example
A	Very high	Poor	Low	Mississippi river (US)
B	Moderate	Good	Very high	James and Mersey river (UK)
C	High	Fair	High	Delaware Bay (US)
D	High	Poor	Moderate	-

(Source: Biggs and Cronin, 1981).

(*) : Following Pritchard's advection-diffusion classification scheme (Pritchard, 1955).

(**): Following Schubel's definition: $I_M = (Q_f T)/(2P_t)$.

In the following sub-sections, firstly, we shall introduce several main classification schemes of estuaries. They include: oceanographic classification, geomorphology classification, tidal classification, stratification classification, classification by salinity curve type and classification of estuaries based on a combination of estuarine characteristics. Finally, multi-channel estuaries will be introduced and classified based on the existing classifications.

2.2.1 Oceanographic classification

Cameron and Pritchard (1963) stated that: " *An estuary is a semi-enclosed coastal body of water which has a free connexion with the open sea and within which sea water is measurably diluted with fresh water from land drainage*". Table 2.2 presents the general characteristics of estuaries as classified by Pritchard (1967).

2.2.2 Classification by tides

There are two common classifications for estuaries based on tidal characteristics. The first classification is based on the tidal range values and the second one is based on the tidal wave propagation characteristics.

Classification by tidal range

The tidal range and freshwater discharge control the type of mixing, circulation and salinity distribution. The tidal range can roughly be used to indicate the type of estuaries. Hayes (1975), who followed the classification proposed by Davies (1964), stated that "*The tidal range has the broadest effect in determining large-scale differences in morphology of sand accumulation*" and that "*a classification of estuaries could best be based on the tidal range*". Table 2.3 presents classification based on the tidal range. However, it does not seem to be a good definition, especially for micro-tidal estuaries, since we can find a number of estuaries having a tidal range smaller than two meters but being partially-mixed or well-mixed (e.g. Limpopo or Gambia).

Classification by tidal wave propagation

The interaction between the tidal wave and the topography of an estuary causes variations in the range of the tide and the strength of the tidal currents. By means of the spatial development of the tidal range, Nichols and Biggs (1985) divided estuaries into hypersynchronous (amplified then damped tidal range), synchronous (un-damped), and hyposynchronous (damped) estuaries. Dyer (1995) indicated that in an ideal estuary, the amount of energy lost by friction is balanced by the amount of energy that is gained by the converging of the riverbanks. This causes the tidal range to be constant along the estuary axis. In an amplified estuary, the tidal range increases in the upstream direction. It is obvious that this process can not continue indefinitely, at some points the friction becomes dominant which leads to a reduction of the tidal amplification and subsequently to tidal damping. In a damped estuary, the friction is larger than the converging of the riverbanks and this leads to a decrease of the tidal range in the upstream direction. Table 2.4 shows the classification based on the tidal wave propagation characteristics. However, it is noted that the word of "synchronous" (i.e. occurring or existing at the same time) seems not to be the right term to define the characteristics of the tidal range. Therefore, in this thesis, we shall use three terms: amplified, un-damped (ideal) and damped estuaries instead of hypersynchronous, synchronous and hyposynchronous estuaries.

Table 2.3 Tidal range classification of estuaries (after Hayes, 1975)

Name	Tidal range (m)	Characteristic	Example
Micro-tidal estuaries	< 2	Mostly highly stratified during high flows	Tampa Bay, Apalachicola Bay, Mississippi (USA), Limfjord, Isefjord (Denmark)
Meso-tidal estuaries	2 - 4	Mostly mixed to partially mixed	Mae Klong (Thailand), Mekong (Vietnam), Lalang (Indonesia), Columbia (USA)
Macro-tidal estuaries	4 - 6	Generally well mixed	Thames, Mersey, Tees (UK), Scheldt (Netherlands), Delaware (USA), Pungue (Mozambique)
Hyper-tidal estuaries	> 6	Generally well mixed	Seine, Somme (France), Severn (UK), Bay of Fundy (Canada)

Table 2.4 Tidal wave propagation classification of estuaries (after Dyer, 1995)

Name	Characteristic	Reason	Example
Amplified (Hypersynchronous) estuaries	Tidal range increases toward the head until the riverine section	Convergence $>$ friction	Scheldt (Netherlands), Seine estuary (France), Humber, Thames (UK)
Ideal (Synchronous) estuaries	Tidal range is almost constant until the riverine section	Convergence $=$ friction	Elbe (Germany), Delaware (UK), Limpopo, Maputo (Mozambique), Gambia
Damped (Hyposynchronous) estuaries	Tidal range decreases toward the estuary head	Convergence $<$ friction	Mekong (Vietnam), Rotterdam Waterway (Netherlands), Incomati, Pungue (Mozambique)

2.2.3 Geomorphology classification

Although many estuarine scientists have used Pritchard's definition, studies in the tidal freshwater regions of estuaries have suggested that the definition of Fairbridge (1980) is also applicable. Fairbridge (1980) stated that: "*An estuary is an inlet of the sea reaching into a river valley as far as the upper limit of tidal rise, usually being divisible into three sectors: a) a marine or lower estuary, in free connections with the open sea; b) a middle estuary subject to strong salt and freshwater mixing; and c) an upper or fluvial estuary, characterized by freshwater but subject to strong tidal action. The limits between these sectors are variable and subject to constant changes in the river discharges* ". Fairbridge proposed estuary classification in seven types (See Table 2.5). The slightly different classification based on geomorphology characteristics can be found in Dyer (1997), namely "Classification by topography"; and Savenije (2005), namely "Classification based on geology".

Table 2.5 Geomorphology classification of estuaries (after Fairbridge, 1980)

Type	Name	Remarks	Example
1			
1a	Fjord	High relief - Shallow sill, constriction in the inlet	Sogne Fjord (Norway), Milford Sound (New Zealand)
1b	Fjard	Low relief – Emerged strandlines	Solway Firth (England/Scotland)
2	Ria	Drowned meanders in the estuary middle sections.	Kingsbridge estuary (UK), Ria de Ribadeo (Portugal), Swan river (Australia)
3	Coastal Plain type – funnel shape	Sea dominant estuary	Chesapeake Bay (USA), Scheldt (Netherlands), Pungue (Mozambique)
4	Bar-built estuary – flask shape	Split bar along coastal line	Vellar estuary (India), Roanoke river (USA)
5	Blind estuary	Ephemeral bar at inlet. Stagnation in dry season.	Balcombe Creek (Australia), Thuan An Inlet (Vietnam)
6	Delta-front estuary	River dominated estuary	Mekong (Vietnam), Nile (Egypt), Mississippi (USA)
7	Tectonic estuary – compound type	Ria (high relief) type at the inlet, Lagoon (low relief) type landward.	San Francisco Bay (USA)

2.2.4 Estuarine stratification classification

Pritchard (1955), Cameron and Pritchard (1963), and later Dyer (1973, 1997) classified estuaries by their stratification and the characteristics of their salinity distributions. This probably is the most common classification for estuaries due to its physical appeal. The advantages of this classification type are to have a better understanding of how the circulation of water in the estuaries is maintained and to get quantification, which should enhance and assist prediction. Four main estuarine types are defined: (i) highly stratified or salt wedge estuaries; (ii) fjords; (iii) partially mixed estuaries; and (iv) well-mixed estuaries (see Table 2.6).

Two points should be emphasised that: (i) a given estuary can be well mixed during the dry period but be partially mixed during high discharge periods; and (ii) a given estuary can consist of several classes, for example it can be well mixed in the lower part and partially mixed in the upper part.

Table 2.6 Stratification classification of estuaries (after Dyer, 1997)

Name	Characteristic	Example
Highly stratified or salt wedge estuaries	Two layers: Upper fresh layer and lower saline layer	Mississippi and Vellar estuary (USA), Mekong (Vietnam – in flood season)
Fjords	Two layers: Fresh upper-intermediate layer and saline deep lower layer	Silver Bay (USA), Alberni inlet (British Columbia)
Partially mixed estuaries	Horizontal and vertical gradually varying density	Rotterdam Waterway (Netherlands), Columbia (USA), Mersey (UK)
Well-mixed estuaries.	Vertical constant density	Mekong (Vietnam – in dry season), Scheldt (Netherlands), Pungue, Incomati, Limpopo (Mozambique), Elbe (Germany)

Quantitative estuary numbers

In the above classifications of estuaries, it can be seen that tide and river discharge are the two dominant drivers for an estuary. Estuary stratification is caused by the density difference between seawater and fresh river water. Kinetic energy supplied by tidal flow reduces the stratification and potential energy supplied by the river discharge enhances stratification. Besides the qualitative classification of estuaries based on their stratification, a number of authors quantitatively classified estuaries on stratification by means of dimensionless numbers, such as Volumetric ratio N, Estuarine Richardson number N_R and Estuary number E_D. Table 2.7 gives a comparison on these three stratification parameters.

The volumetric ratio N, which was introduced by Simmons (1955) and was in fact the Canter-Cremers number, is a ratio between the volume of fresh river discharge coming down the estuary per tidal cycle and the flood volume.

$$N = \frac{Q_f T}{P_t} = \pi \frac{u_f}{\upsilon} \tag{2.1}$$

According to Savenije (1992a): $P_t = EA$ and $E = \frac{\upsilon T}{\pi}$ and hence, $N = \pi \frac{u_f}{\upsilon}$

where Q_f (L^3T^{-1}) is the river flow rate, T (T) is the tidal period, P_t (L^3) is the tidal prism. u_f (LT^{-1}) is the river flow velocity, υ (LT^{-1}) is the tidal amplitude and E (L) is the tidal excursion, which is the distance that a water particle travels between slacks.

Fischer (1979) proposed the "Estuarine Richardson number"(i.e. N_R):

$$N_R = \frac{\Delta\rho}{\rho}\frac{gQ_f}{B\upsilon^3} = \frac{\Delta\rho}{\rho}\frac{gh_0 u_f}{\upsilon^3} = \frac{N}{\pi F_D} \tag{2.2}$$

where B (L) is the estuarine channel width, h_0 (L) is the depth at the estuary mouth. $\Delta\rho$ (MT^{-3}) is the density difference between seawater and river water, ρ (MT^{-3}) is the density of fresh water. g (LT^{-2}) is the acceleration due to gravity and F_D (-) is the estuarine densimetric Froude number: $F_D = \upsilon^2 \Big/ \left(\frac{\Delta\rho}{\rho} g h_0 \right)$.

Thatcher and Harleman (1981) introduced the "Estuary number" E_D:

$$E_D = \frac{1}{\pi} \frac{\rho \upsilon^3}{\Delta\rho g h_0 u_f} = N^{-1} F_D \qquad \text{or} \qquad E_D = \frac{1}{\pi N_R} \tag{2.3}$$

Table 2.7 Classification of estuaries based on stratification parameters

Type of estuaries	N	N_R	E_D
Highly stratified (salt wedge)	$N > 1$	$N_R > 0.8$	$E_D < 0.2$
Partly mixed	$0.1 < N < 1$	$0.08 < N_R < 0.8$	$0.2 < E_D < 8.0$
Well mixed	$N < 0.1$	$N_R < 0.08$	$E_D > 8.0$

2.2.5 Classification by salt intrusion curve shape

For partially mixed and well-mixed estuaries, the classification of Savenije (2005) shows a relation between shape of the salt intrusion curve, geometric shape of an estuary and the hydrology. Accordingly, the following types are distinguished (see Table 2.8 and Fig. 2.1):

(i) Positive estuaries: Recession, Bell and Dome curve shape
(ii) Negative estuaries: Humpback curve shape

Table 2.8 Classification of partially mixed and well-mixed estuaries by salt intrusion curves (after Savenije, 2005)

Type of estuaries	Type of curves	Name	Topography	Example
Positive	1	Recession shape	Straight and narrow, near prismatic estuaries	Limpopo, Incomati (Mozambique), Mekong (Vietnam)
	2	Bell shape	Narrow upstream but strongly funnel-shaped near the mouth.	Pungue (Mozambique), Mae Klong (Thailand), Elbe (Germany)
	3	Dome shape	Wide channels with a pronounced funnel shape	Delaware (USA), Thames (UK), Scheldt (Netherlands)
Negative	4	Humpback shape	Estuaries with rainfall deficit or an evaporation excess	Casamance (West Africa), Laguna Madre (USA)

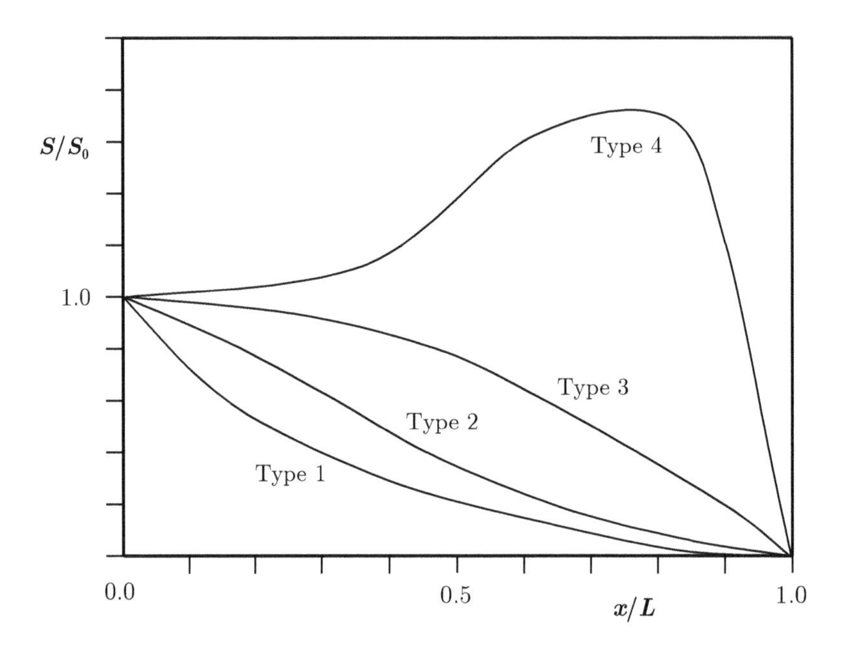

Figure 2.1 Four types of salt intrusion curves (after Savenije, 2005), in which L is the salt intrusion length, x is the distance from the mouth, S_0 is the salinity at the mouth and S is the salinity corresponding with the distance x.

2.2.6 Classification of estuaries based a combination of estuarine characteristics

Besides these classifications, there are other types of classification schemes, for example classifications based on the stratification-circulation diagram (Hansen and Rattray, 1966), morphology (Dalrymple *et al.*, 1992) or river influence (Savenije, 2005). It can be seen that there are many ways to classify estuaries on the basis of their diverse and abundant characteristics. Each classification method is based on one single characteristic or at best, two characteristics of estuaries, with an exceptional case of Cameron and Pritchard, 1963. However, this classification mainly follows the mixing pattern of estuaries.

The question arises whether or not there is an implicit link between those mentioned characteristics, i.e. oceanographic, geomorphology, tidal, stratification characteristics? Savenije (2005) summarized a combined overview of different estuary types based on their main characteristics related to tide, river influence, geology, salinity and stratification (See Table 2.9).

2.2.7 Multi-channel estuaries

How can multi-channel estuaries be classified? A multi-channel estuary, as its name says, has at least two branches. Each branch of the estuary has full hydraulic functions as a single estuarine branch and importantly, each branch interacts with another neighbouring branch.

A number of delta-front estuaries in the "Geomorphology classification" are multi-channel: Mekong Delta (Vietnam), Ganges Delta (Bangladesh), Mississippi Delta (USA) and Yangtze Delta (China), just to name a few. In this kind of multi-channel estuaries, each branch is separated by means of an island, of which we have to take the definition loosely, and each branch can interact with another through junctions (nose and/or toe of the island).

Another kind of multi-channel estuaries can be found in the Western Scheldt in the Netherlands, Pungue (Mozambique), Thames (UK), Columbia (USA) and several estuaries in the Chesapeake Bay (USA). They appear to have a distinct ebb-flood channel system. The ebb and flood channels highly interact through crossover points and small connecting channels. They appear to be in a coastal plain type with a funnel shape in the "Geomorphology classification". The wide funnel shape is an important factor to allow the full development of the ebb and flood channels.

The two main study areas of this thesis, i.e. the Mekong Delta and the Scheldt estuary, are multi-channel estuaries. Looking at Table 2.9, the Mekong Delta corresponds with the estuarine type number 5 and the Scheldt estuary can be classified in the estuarine type number 4. The classification of the study areas is important in order to understand the characteristics of the two systems before carrying out further studies.

Table 2.9 Classification of estuaries based a combination of estuarine characteristics (after Savenije, 2005)

Type	Shape	Tidal wave type	River influence	Geology	Salinity	Estuarine Richardson number
1	Bay	Standing wave	No river discharge	Compound type	Sea salinity	Zero
2	Ria	Mixed wave	Small river discharge	Drowned drainage system	High salinity, often hypersaline	Small
3	Fjord	Mixed wave	Modest river discharge	Drowned glacier valley	Partially mixed to stratified	High
4	Funnel	Mixed wave; large tidal range	Seasonal river discharge	Alluvial in coastal plain	Well-mixed	Low
5	Delta	Mixed wave; small/large tidal range	Seasonal river discharge	Alluvial in coastal plain	Partially mixed to well-mixed	Medium
6	Infinite prismatic channel	Progressive wave	Seasonal river discharge	Man-made	Partially mixed to stratified	High

2.3 MIXING IN ESTUARIES

Although estuaries are diversified due to their unique characteristics, it can be clearly seen in Section 2.2 that the two dominant drivers of an estuary are its tide and river discharge. Moreover, the shape of an estuary certainly defines its own characteristics. The interaction between tide, river discharge (and wind to some extent) and topography causes mixing. Mixing in estuaries is the main reason why a sea and a river can exchange their water, substances and sediments.

Fischer *at al.* (1979) stated that mixing in estuaries results from a combination of small-scale turbulent diffusion and a larger scale variation of the advective mean velocity field. Although mixing in estuaries is much more complex than that in rivers, it can be considered to be a similar physical process, where the main role of diffusion is to transfer mass and momentum between stream lines, and the longitudinal dispersion comes about mainly because the flow along different stream lines is going at different velocities.

Some terms defined by Fischer *at al.* (1979) are used throughout this thesis:

- *"Advection"* represents transport by an imposed current system, as in a river or coastal water.
- *"Dispersion"* is the phenomenon that particles or a cloud of contaminants are scattered by the combined effects of shear and transverse diffusion.
- *"Molecular diffusion"* is the phenomenon that represents the scattering of particles by random molecular motions, which may be described by Fick's law and the classical diffusion equation.
- *"Turbulent diffusion"* is the phenomenon that represents the random scattering of particles by turbulent motion, roughly analogous to molecular diffusion, but with an "eddy" diffusion coefficient.
- *"Mixing"*, in general, is dispersion or diffusion as described above.

There are three main factors which cause mixing and dispersion in estuaries.

- *By tide*: This is probably the most important factor. Tidal flow is considered as a source of kinetic energy. Savenije (2005) identified seven types of mixing due to tidal influence (i.e. tide-driven mixing), namely: (i) turbulent mixing at small spatial and temporal scales; (ii) tidal shear between streamlines with different velocities; (iii) spring-neap tide interaction; (iv) tidal trapping due to trapping of water on tidal flats and in dead ends; (v) residual currents in the cross section; (vi) residual currents over tidal flats and shallows; and (vii) tide-driven mixing due to exchange between ebb and flood channels. The sixth and the seventh types of mixing can be referred as "tidal pumping" (Fischer *et al.*, 1979). Other local effects of tide-driven mixing have also been identified, such as mixing due to tidal amplitude, tidal excursion and junction (Abraham *et al.*, 1986).
- *By river*: The river flow provides potential energy (though buoyancy) that drives density-driven circulation (or gravitational circulation). River discharges determine the volume of freshwater in an estuary and the distribution of the salinity (density) field. They will therefore determine the magnitude of the salinity gradients along the axis of the estuary.

- *By wind*: Wind stresses can generate currents, which affect mixing and dispersion processes. However, with long time-scale mixing such as salinity intrusion, wind effects play a minor role.

In this thesis, we do not pay further attention to wind-driven mixing due to its minor contribution to salt intrusion, the main subject of this study. In the following sections, we shall summarize mixing mechanisms caused by river flow and tides.

2.3.1 Density-driven circulation

The main role of the river flow in mixing is to drive the gravitational circulation in estuaries. Generally, the density-driven circulation (or gravitational circulation) is accompanied by a vertical stratification of salinity. However, in wide estuaries, horizontal (lateral) stratification in general makes the largest contribution to density-driven mixing (Fischer *et al.*, 1979). In the middle part of estuaries, where the salinity gradient is large, the gravitational circulation is supposed to have its largest value. We shall illustrate the density-driven circulation by the vertical stratification that occurs in an estuary with constant width.

Savenije (1993b) developed a simplified schematisation for modelling gravitational circulation (see Fig. 2.2). Gravitational circulation in a well-mixed estuary can be explained as follows:

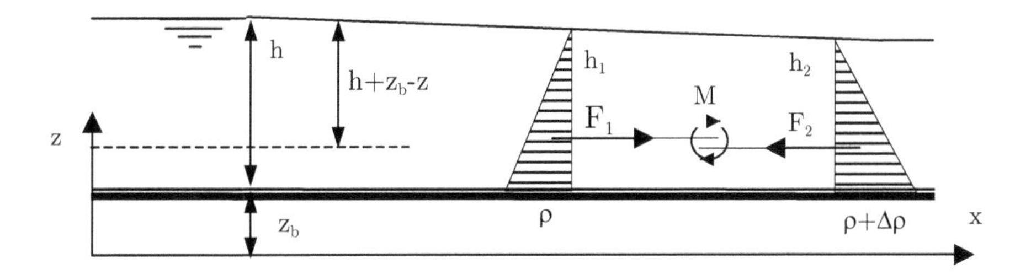

Figure 2.2 Sketch for the gravitational circulation

Considering the unit mass of water, the force per unit mass of water F ($MT^{-2}L^{-2}$) exercised by the water pressure is defined as:

$$F(x,z) = -\frac{1}{\rho}\frac{\partial(\rho g(h + z_b - z))}{\partial x} \tag{2.4}$$

where z (L) is the vertical axis.

According to Van Os and Abraham (1990):

$$F(x,z) = -g\frac{\partial(h + z_b)}{\partial x} - \frac{gh}{2\rho}\frac{\partial\rho}{\partial x} - \frac{g}{\rho}\left(\frac{h}{2} + z_b - z\right)\frac{\partial\rho}{\partial x} \tag{2.5}$$

The third term is z-dependent. At the water surface, where $z = z_b + h$, the third term is equally large as the second term, but of the opposed sign; at the bottom, the third term equals the second term. The second term is independent of z, because $\partial\rho/\partial x$ is assumed not z-dependent in well-mixed estuaries.

Integration over the depth from z_b to z_b+h, division by the depth h yields the depth average water pressure force per unit mass

$$F(x) = -g\frac{\partial(h+z_b)}{\partial x} - \frac{gh}{2\rho}\frac{\partial\rho}{\partial x} - \frac{g}{h\rho}\frac{\partial\rho}{\partial x}\int_{z_b}^{z_b+h}\left(\frac{h}{2}+z_b-z\right)dz \tag{2.6}$$

The first term represents the effect of water surface slope on the net force acting on a unit mass of fluid.

The second term represents the effect of density differences on the instantaneous depth-average tidal flow. This term may be small compared to the first term, but averaged over the tidal period, the second term is larger, and then the first term is negligible.

The third term, of which the depth-averaged value is equal to zero, does not change sign during the tidal cycle. Therefore, it has a dominant effect on the variation of the tidally averaged flow over the depth. It is referred as the "gravitational circulation". Throughout the tidal cycle, from the bottom to mid-depth ($0 < z < ½\,h$), the third term is positive, water is subjected to a landward force; therefore the tidally averaged flow is in the landward direction. Whilst, from the mid-depth to the water surface ($½\,h < z < h$), throughout the tidal cycle, the third term is negative. Hence, water is subjected to a seaward force; therefore the tidally averaged flow is in the seaward direction.

Considering the hydrostatic balance in the unit mass of water (see Fig. 2.2), the two forces F_1 and F_2 that make equilibrium in the horizontal plane over the salt intrusion length L are:

$$F_1 = \frac{1}{2}\rho_1 g h_1^2 \quad \text{and} \quad F_2 = \frac{1}{2}\rho_2 g h_2^2$$

Since $\rho_2 = \rho_1 + \Delta\rho$, there can only be equilibrium if $h_1 > h_2$. However, the two forces, although equal and opposite, exert a momentum that drives the gravitational circulation with an arm equals to $\Delta h/3$.

$$M = \frac{\frac{1}{3}\Delta h \times \frac{1}{2}\rho g h}{\Delta x} = \frac{1}{12}\frac{\partial\rho}{\partial x} g h^2 \tag{2.7}$$

Equation 2.7 implies that the gravitational circulation depends on the longitudinal salinity gradient. This agrees with the finding of Fischer *et al.* (1979) that for wide estuaries, gravitational dispersion is both a function of the width and the salinity gradient. Van de Kreeke and Zimmerman (1990) obtained a perturbation solution for the diffusive regime in relatively narrow estuaries and concluded that the gravitational circulation is proportional to the longitudinal salinity gradient and the depth to the power three. McCarthy (1993) using a 2D vertical model and perturbation analysis confirmed that density-driven dispersion is weak at the estuary mouth of a wide estuary and it is strong further inland as a result of the large salinity gradient. McCarthy also found that density-driven dispersion is a function of the salinity gradient.

The above finding is somewhat contrary to results of several studies about the gravitational circulation in estuaries. Smith (1980) found that the density-driven dispersion is dominant in wide estuaries. West and Broyd (1981) confirmed that tide-driven shear mechanisms dominated for narrow, shallow estuaries, while the density-driven shear mechanisms dominated in wide estuaries. An explanation for this contradiction can be found in Savenije (2005, pp. 114-115). The contradiction can be overcome if we distinguish between width and convergence. An estuary with a short convergence length is wide at the mouth. An estuary with a long convergence length is considered narrow. Strongly converging estuaries are dominated by tide-driven mixing while weakly converging (near constant cross-section) estuaries are dominated by density-driven mixing.

2.3.2 Tide-driven circulation

There are a number of mixing mechanisms caused by tides as indicated earlier. Turbulent mixing is in fact the weakest of the mixing mechanisms occurring only at small spatial and temporal scales. Spring-neap tide interaction, residual currents in the cross-section, mixing due to tidal amplitude, tidal excursion and junction could be important for certain estuaries. However, they do not significantly contribute to the steady-state longitudinal salt dispersion, which is the main subject of this thesis.

Tidal trapping, which is the concept introduced by Schijf and Schönfeld (1953), is due to the phase difference between the main estuary branch and a dead-end tidal branch, bay or tidal flat. They pointed out how phase lags between the currents in a shallow embayment and the currents in the adjoining main tidal channel can cause additional mixing. As the tide enters the estuary, it fills tidal flats and bank irregularities. On the ebb tide, the emptying of these pockets of relatively saline water is generally delayed, resulting into the longitudinal dispersion of salt. Okubo (1973) gave an analysis to investigate the trapping mechanism in estuaries and other embankments. In the Mersey (UK), which is an estuary with highly irregular topography, it appeared that the trapping mechanism itself could account for the longitudinal effective dispersion.

Fischer *et al.* (1979) defined "tidal pumping" as the energy available in the tide that drives steady circulations similar to what would happen if pumps and pipes were installed to move water about in circuits. Tidal pumping is an important large-scale mixing mechanism for moving pollutants and transporting salinity upstream against a mean outflow of fresh water. There are two types of residual circulation that cause tidal pumping: (i) interaction of the tidal flow with a pronounced flood-ebb channel system; and (ii) interactions of the tidal flow with the irregular bathymetry, which in fact is the tidal trapping mechanism defined earlier. Residual ebb-flood channel circulation is an important large-scale mixing mechanism for moving pollutants and transporting salinity upstream against a mean outflow of fresh water as shown in the Scheldt estuary (Van Veen, 1950; Van Veen *et al.*, 2005; Jeuken, 2000; and Savenije, 2005). Abraham *et al.* (1975) also noted that: "*The contribution of transverse variations is certainly significant in wide estuaries having ebb channels different from flood channels, and tidal flats*".

We shall introduce some approaches to investigate and to quantify the tide-driven mechanism. They can be grouped as "Decomposition method" and "Other methods" in the following sub-sections.

Decomposition method

A vast number of authors have tried to quantify the tidal pumping mechanisms (Bowden, 1963; Hansen, 1965; Fischer, 1972; Lewis, 1979; Uncles and Jordan, 1979; Uncles *et al.*, 1985; Pino Q *et al.*, 1994; or Sylaios and Boxall, 1998). They mainly employed a decomposition method to estimate tidal pumping in terms of salt fluxes in order to evaluate the relative importance of tidal pumping in different estuaries compared to other mechanisms such as gravitational circulation, tidal trapping, tidal straining etc. The decomposition method, therefore, serves the purpose of determining not only the tidal pumping but also other mechanisms.

Below we present the decomposition method of Uncles *et al.* (1985) to estimate tidal pumping and gravitational circulation over a cross-section.

The cross-sectional instantaneous rate of water transport

The instantaneous rate of water transport Q (L^3T^{-1}) through a cross-section A (L^2), is given by:

$$Q = A\overline{u} \tag{2.8}$$

where u (LT^{-1}) is the instantaneous velocity and the overbar denotes an average over the cross section.

The tidal average (residual) water transport is defined as:

$$\langle Q \rangle = \langle A \rangle [V_1 + V_2] \tag{2.9}$$

in which the $\langle\ \rangle$ brackets denote the tidal average of a variable.

V_1 (LT^{-1}) is the cross-sectional averaged Eulerian residual current (or non-tidal drift current):

$$V_1 = \langle \overline{u} \rangle = \frac{1}{T}\int_0^T \left(\frac{1}{A}\int_0^A u \mathrm{d}A \right) dt \tag{2.10}$$

If we subdivide the cross-section into discrete cells, then Eq. 2.10 is approximated by:

$$V_1 \approx \frac{1}{T}\sum_{k=1}^{t} \left(\frac{1}{A}\sum_{i=1}^{n}\sum_{j=1}^{m} A_{ij} u_{ij} \right) \Delta t \tag{2.11a}$$

where m is the number of columns, n is the number of rows. Δt and t are the sub-timestep and the number of sub-timesteps within one tidal period. T (T) is the tidal period. A_{ij} (L^2) and u_{ij} (LT^{-1}) are the area and velocity of the discrete cells.

V_2 (LT^{-1}) is the Stokes drift current, defined as:

$$V_2 = \left\langle \tilde{A}\tilde{u} \right\rangle / \langle A \rangle \tag{2.11}$$

in which $\tilde{u} = \overline{u} - \langle \overline{u} \rangle$ and $\tilde{A} = A - \langle A \rangle$.

$V_L = [V_1 + V_2]$ is the Lagrangean current, which in the absence of evaporation and rainfall corresponds to the tidal average velocity of the fresh water discharge that flows through the cross-section.

The cross-sectional instantaneous rate of salt transport

The instantaneous rate of salt transport Q_S (L^3T^{-1}) through a cross section A, is given by:

$$Q_S = A\overline{uS} \tag{2.12}$$

where S (-) is the salinity. Furthermore, the residual transport rate of salt is in the form:

$$\langle Q_S \rangle = \langle A \rangle \left[V_{S,1} + V_{S,2} + V_{S,3} \right] + V_S^* \tag{2.13}$$

in which the four transport components are defined below.

$V_{S,1}$ (LT^{-1}) is the cross-sectional residual flux of salt due to the residual transport of water:

$$V_{S,1} = \langle Q \rangle \langle \overline{S} \rangle / \langle A \rangle \tag{2.14}$$

where:

$$\langle \overline{S} \rangle = \frac{1}{T} \int_0^T \left(\frac{1}{A} \int_0^A S \mathrm{dA} \right) dt \tag{2.15}$$

If we subdivide the cross-section into discrete cells, then Eq. 2.15 is approximated by:

$$\langle \overline{S} \rangle \approx \frac{1}{T} \sum_{k=1}^{t} \left(\frac{1}{A} \sum_{i=1}^{n} \sum_{j=1}^{m} A_{ij} S_{ij} \right) \Delta t \tag{2.15a}$$

where S_{ij} (-) is the salinity of the discrete cells.

Combination of Eqs. 2.9 and 2.14 yields:

$$V_{S,1} = [V_1 + V_2] \langle \overline{S} \rangle = V_L \langle \overline{S} \rangle \tag{2.16}$$

$V_{S,2}$ (LT^{-1}) represents the cross sectional averaged residual flux of salt due to the non-zero temporal correlation between $\tilde{Q}$ (L^3T^{-1}) and $\tilde{S}$ (-), which is called the tidal pumping:

$$V_{S,2} = \langle \tilde{Q} \tilde{S} \rangle / \langle A \rangle \tag{2.17}$$

in which $\tilde{Q} = Q - \langle Q \rangle$ and $\tilde{S} = \overline{S} - \langle \overline{S} \rangle$.

$V_{S,3}$ (LT^{-1}) represents the cross sectional averaged residual flux of salt due to the non-zero spatial correlation or shear between tidal and residual currents. This mechanism is driven by density gradients and is called gravitational circulation:

$$V_{S,3} = \langle A \overline{u' S'} \rangle / \langle A \rangle \tag{2.18}$$

where u' (LT^{-1}) and S' (-) are the deviations of velocity and salinity from the cross-sectional value, respectively.

By definition, $\langle \tilde{Q} \rangle = \langle \tilde{A} \rangle = \langle \tilde{u} \rangle = \langle \tilde{S} \rangle = 0$ and $\overline{u'} = \overline{S'} = 0$.

V_S^* is due to "interact" processes. V_S^* is found to be small compared to $V_{S,1}$, $V_{S,2}$ and $V_{S,3}$ (Uncles *et al.*, 1985; and Pino Q *et al.*, 1994). Therefore V_S^* can be neglected.

Many authors have followed the decomposition method and they have obtained a number of important components for mixing mechanisms, ranging from the most important three mechanisms (i.e. residual flux of water, residual flux of tidal pumping and residual flux of mean shear effect produced by gravitational circulation) in Uncles and Jordan (1979), Uncles *et al.* (1985), Pino Q *et al.* (1994) or Sylaios and Boxall

(1998); six components (i.e. advection, geometry-induced dispersion, residual lateral circulation, vertical density circulation, lateral oscillatory shear and vertical shear) in van de Kreeke and Zimmerman (1990) ; eight components (Winterwerp, 1983); or even 11 components (Park and James, 1990).

Although the decomposition method has significantly contributed to the understanding of estuarine mixing processes, especially to investigate dominant mechanisms in a particular estuary, the method has a number of problems (Rattray and Dworski, 1980; Geyer and Signell, 1992; Jay *et al.*, 1997; and Savenije, 2005).

- It is highly data intensive to determine a residual flux in a cross section. To get the adequate data set, one has to set up a complex monitoring campaign, which has to be done during several tidal periods and sample the entire cross section at many width and depth points. Moreover, substantial errors on measurements may result (Lane *et al.*, 1997; and Lewis, 1997).
- Mixing in estuaries is a 3-D process that acts mainly in the longitudinal direction. Therefore, a large number of cross sections should be monitored in order to have sufficient information. Considering the already large problem to obtain data for a single cross-section, it would not be feasible to get data for a large number of cross sections.
- The error made in determining residual transport fluxes can be substantial. In the case of tidal hydraulics, the momentary fluxes are much larger than the residual fluxes, which are in fact obtained by averaging the momentary fluxes over one tidal cycle. Thus, the errors in determining the residual fluxes can easily exceed the residual fluxes themselves.
- The close interrelation between different decomposed mixing components poses a doubt in analysing the contribution of each individual component. On the one hand, the separation between tidal pumping and gravitational circulation is obvious due to the different natures of mixing. On the other hand, the separation between vertical and transverse (i.e. lateral) gravitational circulation is much dependent on the way that one chooses the detail of the decomposition on velocity and salinity fields. Moreover, it is hard to distinguish between lateral dispersion and advection in an estuary since the transverse transport of large scales such as secondary circulation cannot be described simply as a dispersion process.

Nevertheless, with sufficient data, the decomposition method is a useful tool to investigate dominant mixing mechanisms in estuaries, at least between the two main mechanisms: tidal pumping and gravitational circulation (i.e. tide-driven and density-driven circulation). The development of 2-D and 3-D hydrodynamic models, especially with the 3-D hydrodynamic models, can provide vital 3-D detailed data to investigate the mixing mechanisms in a "virtual laboratory". The models, of course, have to be treated with care in order to reproduce the correct velocity field and salinity field.

Other methods

Besides the decomposition method, several approaches have been proposed for quantifying tide-driven dispersion. The tidal prism approach was first introduced by

Ketchum (1951) and later adopted by Arons and Stommel (1951) in their so-called mixing-length theory. Arons and Stommel formulated a tidal dispersion coefficient as follows:

$$D = k_A U_0 l_0 \tag{2.19}$$

where D (L^2T^{-1}), k_A (-), U_0 (LT^{-1}) and l_0 (L) are the dispersion coefficient, a constant, velocity scale and mixing length scale, respectively. k_A is understood to be a constant for one estuary. However, because the values of k_A widely varied, it resulted in a very large range of computed values and therefore Eq. 2.19 could not be used for predictive purposes.

Zimmerman (1976) developed the "tidal random walk" theory, which considers the Lagrangean motions in estuaries resulting from the purely advective effects of tidal and residual currents and takes into account pronounced horizontal residual circulations generated by tide-topography interactions. Due to the presence of residual eddies, the velocity field cannot be considered uniform and in fact it varies considerably over distances in the order of the tidal excursion E. Zimmerman (1976, 1981) derived an equation for the longitudinal dispersion coefficient as:

$$D = k_A(U_0, l_0) U_0 l_0 \tag{2.20}$$

Equation 2.20 is formally the same as the equation of Arons and Stommel. The main innovation in the equation of Zimmerman is that k_A has a well-defined physical meaning, being a function of U_0 and l_0 through the dimensionless parameters reflecting mixing length and tidal velocity. Zimmerman (1976) successfully predicted a value of D that agreed with estimates of dispersion obtained from the salt budget of the Dutch Wadden Sea.

Besides these two approaches, Geyer and Signell (1992) stated that the development of numerical models (i.e. 2-D and 3-D hydrodynamic models) provides a mean of isolating the tide-driven circulation and dispersion from the other processes, and they can simulate the non-linear effects that produce tide-driven circulations (Nihoul and Ronday, 1975; and Tee, 1978) and transient eddies (Imasato, 1983), both of which may contribute to tide-driven dispersion. Moreover, taking the advantage of numerical models, remote sensing and high-tech measurements, a particle tracking technique was used to investigate the influence of tide-driven dispersion in estuaries (e.g. Wolanski and Heron, 1984; or Signell and Geyer, 1990).

2.4 SALT TRANSPORT AND SALINITY INTRUSION MODELS IN ESTUARIES

The first part of this section presents a summary on the development of the salt balance equation, which is a tool for forecasting salinity movements in estuaries. The salt balance equation has appeared in many publications (e.g. Fischer *et al.*, 1979; Dyer, 1997; or Savenije, 2005), however it is worthwhile to summarize here in order to have a good overview. The second part of this section introduces several methods to predict salinity intrusion and the effective salt dispersion coefficient.

2.4.1 Salt balance equation

In order to adequately estimate and predict salt transport characteristics in estuaries, it is important to find a way to quantify the water circulation and the mixing processes. It can be done by considering the budget of salt within sections of one estuary and assuming the conservative property of salt. Since this study mainly focuses on the 1-D process of salinity movement, the salt balance equation will be introduced in a 1-D form.

Fick's law states that the dispersive flux of solute mass (the mass of a solute crossing a unit area per time in a given direction) is proportional to the gradient of solute concentration in that direction. Salt dispersion in estuaries can be described as an analogy of Fick's law. In three dimensions, it can be written as follows:

$$F = D\nabla s \tag{2.21}$$

where F ($ML^{-1}T^{-1}$) is the solute mass flux vector per unit width with components (F_x, F_y, F_z). s (ML^{-3}) is the mass concentration of diffusing solute and D (L^2T^{-1}) is the coefficient of proportionality (i.e. the diffusion coefficient).

In one dimension, this equation can be rewritten as:

$$F_x = D_x \frac{\partial s}{\partial x} \tag{2.22}$$

in which x (L) is the longitudinal axis.

We now consider the conservation of mass in a one-dimensional transport process. In the following 1-D description, we shall remove the subscript x for reasons of simplicity.

The total rate of mass transport is the diffusive flux ($-D\frac{\partial s}{\partial x}$) plus the advective flux (us):

$$F = us + \left(-D\frac{\partial s}{\partial x}\right) \tag{2.23}$$

The mass-conservation equation reads:

$$\frac{\partial s}{\partial t} + \frac{\partial F}{\partial x} = 0 \tag{2.24}$$

From Eq. 2.23: $$\frac{\partial F}{\partial x} = -\frac{\partial}{\partial x}\left(us + \left(-D\frac{\partial s}{\partial x}\right)\right)$$

Combining this with Eq. 2.24 yields: $$\frac{\partial s}{\partial t} + \frac{\partial (us)}{\partial x} - \frac{\partial}{\partial x}\left(D\frac{\partial s}{\partial x}\right) = 0 \tag{2.25}$$

Equation 2.25 is called the advection-diffusion equation in a one-dimensional transport process for substances. If salt is the considered substance, then Eq. 2.25 is called the salt balance equation.

In Savenije (2005), a cross-sectional average equation based on a realistic tidal hydraulic form of the salt balance, which takes into account the different states of water discharges, estuarine sections and storage width ratios, was presented:

$$r_s A\frac{\partial s}{\partial t} + Q\frac{\partial s}{\partial x} - \frac{\partial}{\partial x}\left(AD\frac{\partial s}{\partial x}\right) = -sR_S \tag{2.26}$$

where A (L^2) is the cross section area and Q (L^3T^{-1}) is the discharge. r_s (-) is the storage width ratio and R_S (L^2T^{-1}) is the source term.

The storage width ratio can be disregarded in steady state models, however it should be included in unsteady-state models. The R_S term is often disregarded in the salt balance equation when computing for positive estuaries with minor rainfall, evaporation or lateral inflows. However for the case of negative estuaries (see Section 2.2.5) or estuaries with considerable amounts of rainfall, evaporation and/or lateral flows, this term can play an important role. Some examples for the importance of R_S can be seen in Savenije (1988) and Savenije and Pagès (1992).

2.4.2 Salt transport and salt intrusion models

To obtain solutions for the salt balance equation (i.e. Eq. 2.26), one has to know the values of discharge, area, source term, storage width ratio, and most importantly, the dispersion coefficient. The first four values can be obtained by solving the set of equations for conservation of mass and conservation of momentum. The latter (i.e. the effective dispersion coefficient) incorporates all the dispersion mechanisms that counteract the advective salt transport. It can be found mainly based on prototype and observed salinity characteristics of the considered estuary. Alternatively, if a predictive theory exists, the dispersion can be computed on the basis of hydraulic and topographic parameters (as in Savenije, 2005).

Generally, there are two approaches to solve the salt balance equation, namely steady-state and unsteady-state models. The use of steady-state models gives an opportunity to develop predictive models for salinity intrusion in estuaries. In the following sections, we shall briefly introduce them.

Steady-state models

Many authors solved the salt balance equation for a state of (quasi-) equilibrium in an estuary (Ippen and Harleman, 1961; van Dam and Schönfeld, 1967; Van der Burg, 1972; Savenije, 1989, 1993c; Lewis and Uncles, 2003; Prandle, 2004; and Brockway *et al.*, 2006). For a long time, this equilibrium-state (i.e. steady-state) approach was the only one that would yield practical results for forecasting salinity intrusion.

The steady-state models proceed from the salt balance equation with the assumption that the discharge and the rate of change of the salinity can be decomposed into a tidal component with a periodicity that equals the tidal period and a long-term component. Assumptions made depend on the moments of consideration, i.e. tidal average situation (TA) or high water slack (HWS) or low water slack (LWS). We shall briefly introduce two models to demonstrate procedures to obtain steady-state salt balance models. The first one is from Ippen and Harleman (1961) for LWS and the second one is from Savenije (2005) for LWS, HWS and TA situations.

LWS model

Ippen and Harleman (1961) considered conditions at LWS, assuming that:

$\frac{\partial s}{\partial t} = 0$ and $u_t = 0$

Considering estuaries with constant cross section, Eq. 2.25 can be rewritten as:

$$-u_f \frac{\mathrm{d}s}{\mathrm{d}x} = \frac{\mathrm{d}}{\mathrm{d}x}\left[D_x^{LWS} \frac{\mathrm{d}s}{\mathrm{d}x} \right] \quad (2.27)$$

where u_t and u_f (LT^{-1}) are the tidal component and the long-term component of the velocity u, respectively. D_x^{LWS} is the LWS dispersion coefficient at location x from the mouth.

Integrating Eq. 2.27, assuming u_f is constant in a prismatic channel, yields:

$$-u_f s = D_x^{LWS} \frac{ds}{dx} \quad (2.28)$$

From the results of laboratory experiments, Ippen and Harleman proposed that:

$$D_x^{LWS} = \frac{D_0^{LWS} \times X_S}{X_S + B} \quad (2.29)$$

in which X_S (L) is the distance from the river mouth to the point where the salt content is equal to that of seawater. D_0^{LWS} is the LWS dispersion coefficient at the mouth.

Integrating with respect to x and substituting Eq. 2.29 into Eq. 2.28 yield:

$$\frac{s}{s_z} = \exp\left[-\frac{u_f}{2D_0^{LWS} X_S}(x + X_S)^2 \right] \quad (2.30)$$

where s_z (-) is the salinity content of seawater and the origin of x is located at the estuary mouth.

Harleman and Abraham (1966) found the following relation between D_0^{LWS} and X_S:

$$\frac{D_0^{LWS}}{u_f X_S} = 0.055 \left(\frac{h_o}{\eta} \right)^{2.7} E_D^{\ 1.2} \quad (2.31)$$

where η (L) is the tidal amplitude and h_0 (L) is the depth at the river mouth.

It is observed that this LWS model assumes a constant cross-section; therefore it does not take into account the longitudinal variation of the estuarine cross-section.

LWS, HWS and TA models

Savenije (2005) introduced three forms for LWS, HWS and TA models. We only introduce the TA situation in order to demonstrate the procedures for the steady-state models. The HWS and LWS steady-state models can be obtained through the same procedures, although they have some slightly different assumptions.

Savenije (2005) considered the tidal average situation, which follows from averaging over a tidal cycle under several first order approximations:

$$\frac{1}{T}\int_0^T A\frac{\partial s}{\partial t}\mathrm{d}t \approx A_{TA}\frac{\partial s_{TA}}{\partial t} \tag{2.32}$$

$$\frac{1}{T}\int_0^T Q_t\frac{\partial s}{\partial x}\mathrm{d}t \approx 0 \tag{2.33}$$

$$\frac{1}{T}\int_0^T Q_f\frac{\partial s}{\partial x}\mathrm{d}t \approx Q_f\frac{\partial s_{TA}}{\partial x} \tag{2.34}$$

$$\frac{1}{T}\int_0^T \frac{\partial}{\partial x}\left(AD\frac{\partial s}{\partial x}\right)\mathrm{d}t \approx \frac{\partial}{\partial x}\left(A_{TA}D_{TA}\frac{\partial s_{TA}}{\partial x}\right) \tag{2.35}$$

in which the subscript TA refers to the tidal average situation, and D_{TA} is the tidal and cross-sectional average, x-dependent, dispersion. Q_t and Q_f are the tidal component and the long-term component (i.e. freshwater discharge) of the discharge Q.

Making use of the above assumptions and ignoring the source term, Eq. 2.26 can be rewritten as:

$$r_S A_{TA}\frac{\partial s_{TA}}{\partial t} + Q_f\frac{\partial s_{TA}}{\partial x} - \frac{\partial}{\partial x}\left(A_{TA}D_{TA}\frac{\partial s_{TA}}{\partial x}\right) = 0 \tag{2.36}$$

The weakness of the above assumptions and limitations on the use of Eq. 2.36 can be found in Fischer *et al.* (1979) and Savenije (2005).

If a steady state occurs, implying that $\partial s_{TA}/\partial t = 0$, then Eq. 2.36 can be modified into:

$$Q_f\left(S_{TA} - S_f\right) - A_{TA}D_{TA}\frac{\mathrm{d}S_{TA}}{\mathrm{d}x} = 0 \tag{2.37}$$

where S_{TA} represents the mean tidal steady-state salinity. The boundary condition used is that $S_{TA}=S_f$ and $\partial S_{TA}/\partial x = 0$ when $x\to\infty$. The river water salinity S_f (-) is generally small as compared to the salinity in the estuary ($S_{TA} >> S_f$) and is often disregarded.

Due to the similarity between the steady-state equations for the TA, HWS and LWS situations, they can be combined in a general form:

$$S_i - S_f = \frac{A_i D_i}{Q_f}\frac{\mathrm{d}S_i}{\mathrm{d}x} \tag{2.38}$$

in which S_i (-) indicates the salinity of the three different states: HWS, LWS and TA

Considering the exponential shape of estuaries, i.e. $A = A_0\exp(-x/a)$, wherein A_0 (L^2) is the tidal average cross-sectional area at the estuary mouth; a (L) is the area convergence length and x (L) is the distance from the mouth; and using the expression of Van den Burgh (1972) (see Section 3.2), Eq. 2.38 can be integrated, yielding:

$$\frac{S_i - S_f}{S_{i0} - S_f} = \left(1 + \frac{KaQ_f}{D_{i0}A_0}\left(\exp\left(\frac{x}{a}\right) - 1\right)\right)^{\frac{1}{K}} \tag{2.39}$$

where K (-) is the Van den Burgh's coefficient. D_{i0} (L^2T^{-1}) and S_{i0} (-) are the boundary conditions at the river mouth (x=0) for the HWS, LWS or TA conditions. The dispersion coefficient D_{i0} is determined based on salinity observations of the considered situation or based on a predictive equation. The details on this model shall be further introduced in Chapter 3.

State of equilibrium

Steady-state models have as the most important assumption that the estuary is in a state of equilibrium. A steady state model, which does not allow temporal variation, on the one hand offers practical and simple solutions, on the other hand has a number of disadvantages (Blumenthal *et al.*, 1976).

(i) Estuaries do not always reach such an equilibrium, mainly because a long time is needed before an equilibrium state occurs in nature and because the boundary conditions of tide and river outflow vary in time.

(ii) The dispersion coefficient is different for different averaging periods.

The time required for an equilibrium to occur depends on:(i) the rate at which the boundary conditions vary, in particular the rate of change of the fresh water discharge; and (ii) the time required for the system to adjust itself to a new situation.

To arrive at a steady state equation for conservation of mass, it is required that in the estuary an equilibrium condition is reached between, on the one hand, advective salt transport through the downstream flushing of salt by the fresh water discharge, and on the other hand, the full range of mixing processes. Savenije (2005) proposed an expression derived for the system response time as a function of the steady state salinity distribution in order to investigate how quickly an estuary system adjusts to a new situation. If the time required for the system to adjust is not too long in relation to the variation of the boundary conditions, then a steady state model may be used. The expression reads:

$$T_s \approx -\frac{1}{Q_f S(L/2)} \int_{L/2}^{L} ASdx \tag{2.40}$$

where T_s (T) is the system response time, which represents the time required for the system to adjust itself from one steady state to another and S (-) is the steady-state salinity. L (L) is the salt intrusion length.

Generally, the estuary reacts relatively quickly to an increase of the river discharge, whilst the reaction to a decrease in the fresh water discharge is slow, since the process of salinization requires gradual replacement of the fresh water by saline water through mixing. Savenije (2005, p. 175) found that the system response times for different estuaries are in the order of magnitude of days to months, e.g. Mae Klong (1 day), Pungue (25 days); Scheldt (30 days); Incomati (36 days) or Gambia (2270 days). Nguyen and Savenije (2006) found that the system response time during the dry season in the Mekong Delta is in the order of one week.

Predictive models

The use of steady-state models provides an opportunity to develop predictive models for salinity intrusion in estuaries. In a predictive model, the dispersion coefficient is determined as a function of hydrologic, hydraulic and topographic parameters, which are measurable or quantifiable variables. The function is often semi-empirical and is established based on the calibration results of steady-state models against prototype data or observations.

The salt intrusion length, which is the distance from the estuary mouth to the point where the salinity reaches the river salinity, is the most important output of a predictive model. There are three types of intrusion length: intrusion length at low water slack (L^{LWS}), intrusion length at high water slack (L^{HWS}) and tidal average intrusion length (L^{TA}), which is considered to be an average of L^{LWS} and L^{HWS}. There are a number of available predictive models: Van den Burgh (1972), Rigter (1973), Fischer (1974), Van Os and Abraham (1990), Savenije (1993, 2005) and Prandle (2004).

Van den Burgh (1972) using the prototype data of Dutch estuaries arrived at the following equation:

$$L^{TA} = -\frac{26h_0}{K}\frac{\sqrt{gh_0}}{\upsilon_0}\frac{\upsilon_0}{u_0}N^{0.5} = 26\pi\frac{h_0}{K}F^{-1.0}N^{-0.5} \tag{2.41}$$

where υ_0 (LT^{-1}) is the tidal velocity amplitude at the river mouth. u_0 (LT^{-1}) is the fresh water velocity at the estuary mouth: $u_0 = Q_f / A_0$ where A_0 is the cross sectional area at the estuary mouth. Finally, F (-) is the Froude number: $F = \left(u_0^2\right)/\left(gh_0\right)$. Other notations have already been presented in Section 2.2.4.

Rigter (1973), based on flume data of Delft Hydraulics Laboratory (DHL) and of the Waterways Experiment Station (WES), proposed:

$$L^{LWS} = 1.5\pi\frac{h_0}{f}(F_D^{-1}N^{-1} - 1.7) \approx 4.7\frac{h_0}{f}F_D^{-1}N^{-1} \tag{2.42}$$

where f (-) is Darcy-Weisbach's roughness.

Fischer (1974) also used the data of DHL and WES and proposed:

$$L^{LWS} = 17.7\frac{h_0}{f^{0.625}}F_D^{-0.75}N^{-0.25} \tag{2.43}$$

Van Os and Abraham (1990) developed an equation, which is very close to the equation of Rigter:

$$L^{LWS} = 4.4\frac{h_0}{f}F_D^{-1}N^{-1} \tag{2.44}$$

Savenije (1993c) based on Eq. 2.39 proposed a new formula for predictive salt intrusion that takes into account estuary geometry. The equation is written as:

$$L^{HWS} = a\ln\left(-220\frac{h_0 E_0 \upsilon_0}{Ka^2 u_0}F^{-0.5}N^{0.5}+1\right) \tag{2.45}$$

where E_0 (L) is the tidal excursion at the mouth.

Equation (2.45) was slightly revised in Savenije (2005) into:

$$L^{HWS} = a\ln\left(-1400\frac{h_0 E_0 \upsilon_0}{Ka^2 u_0}N_R^{0.5}+1\right) \tag{2.46}$$

In Chapter 3 and Chapter 4 of this thesis, the predictive salinity model of Savenije will be further analysed.

Prandle (2004) proposed the following equation to compute L^{TA}, which is restricted to partially mixed estuaries and basing on the tidally linearised theories relating to the vertical structure of salinity and velocity:

$$L^{TA} = \frac{5h_0^3 B_0 X_c^{3m+n}}{Q_f k \upsilon_0} \tag{2.47}$$

where k (-) is the bed friction. h_0 (L) and B_0 (L) are the depth and width at the estuary mouth. m (-) and n (-) are the power coefficients of width and depth variation. X_c (L) is the central location.

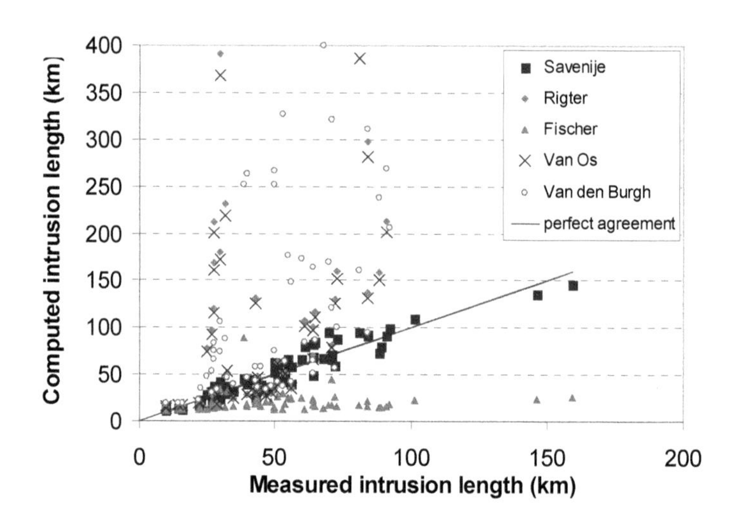

Figure 2.3 Salinity intrusion length at HWS computed by several methods (after Savenije, 2005)

All the above-mentioned models were derived for single-channel estuaries. It is remarked that the models of Rigter (1973), Fischer (1974) and Van Os and Abraham (1990) were developed under an assumption of constant cross section. The major advantage of the model of Savenije (1993c, 2005) is that this model takes into account the exponential shape of estuaries, i.e. $A = A_0 \exp(-x/a)$ and $B = B_0 \exp(-x/b)$, wherein A_0 (L^2), B_0 (L) are the area and width at the estuary mouth; a, b (L) are the convergence lengths of area and width; and x (L) is the distance from the mouth. Prandle (2004) described the bathymetry as a power function: $B = B_0 x^n$ and $h = h_0 x^m$. However, this bathymetric approximation is considered unrealistic. As one can see in Fig. 2.3, the results of Savenije's model so far is considered most accurate.

Unsteady-state models

In estuaries, where a steady state hardly exists such as in the Gambia, unsteady-state models are the only useful tools to simulate and predict salinity movement. The unsteady-state models provide simultaneously a solution for the set of required equations (i.e. conservation of mass, conservation of momentum, salt balance and state equations). The set of equations for 1-D unsteady-state models is briefly introduced below:

$$\frac{\partial Q}{\partial t} + \alpha_S \frac{\partial (Q^2/A)}{\partial x} + gA\frac{\partial h}{\partial x} + gA\frac{\partial Z_b}{\partial x} + gA\frac{h}{2\rho}\frac{\partial \rho}{\partial x} + gA\frac{U|U|}{C^2 h} = 0 \tag{2.48}$$

$$r_S \frac{\partial A}{\partial t} + \frac{\partial Q}{\partial x} = 0 \tag{2.49}$$

The salt balance equation (Eq. 2.26):

$$r_s A\frac{\partial s}{\partial t} + Q\frac{\partial s}{\partial x} - \frac{\partial}{\partial x}\left(AD\frac{\partial s}{\partial x}\right) = -sR_S$$

And the state equation:

$$\rho = 0.75s + 1000 \tag{2.50}$$

where α_S (-) is the shape factor (assumed constant) to account for the spatial variation of the flow velocity over the cross-section ($\alpha_S > 1$); $Z_b = Z_b(x,t)$ (L) is the mean cross-sectional bottom elevation; $U = U(x,t)$ (LT^{-1}) is the mean cross-sectional flow velocity; $C = C(x)$ ($L^{0.5}T^{-1}$) is the coefficient of Chezy; $h = h(x,t)$ (L) is the mean cross-sectional depth of flow. Other notations have already been introduced earlier.

For a detailed view on this set of equations, the reader is referred to Savenije (2005). To date, a vast number of unsteady-state models have been developed, such as Stigter and Siemons (1967), Thatcher and Harleman (1981), Prandle (1981) and Savenije (2005).

The unsteady-state models demand a numerical approach, since the set of required equations cannot be solved analytically. A number of numerical approaches have been developed, mainly based on the finite difference approach (Cunge *et al.*, 1980; Abbott and Basco, 1989; Gross *et al.*, 1999; or Lin and Falconer, 1997). A disadvantage of all 1-D models is that they use the dispersion D as a calibration coefficient, without making use of a model that prescribes the dispersion as a function of geometry, discharge and tide. This reduces the predictive capacity of these models. The only model that uses a predictive equation for D is the one by Savenije (2005). Besides 1-D unsteady-state models, 2-D and 3-D unsteady-state models have been developed. Recently, several numerical models have become available, such as SOBEK-RIVER (1-D), SOBEK-

RURAL (1-D and 2-D), HES-RAS (1-D), ISIS (1-D), MIKE11 (1-D), MIKE21 (2-D), MARS-2D (2-D), MIKE3 (3-D) or DELFT3D (3-D). These models allow us to investigate the hydrodynamic and salt transport regimes of an estuary with integrated GIS friendly-user interface and a wide application range.

It is remarked that the set of required equations needs boundary and initial conditions for the tidal motion, river flow as well as salinity. For 1-D hydrodynamic computations, the tidal water level as a function of time is used as the downstream boundary. For the upstream boundary, the discharge is generally used.

For salinity conditions, the upstream boundary condition is set as $\partial s / \partial x = 0$ for both closed and open-end estuaries. Normally, the upstream salinity condition can be set equal to the river salinity, since the upstream boundary is outside the zone of salt intrusion. The downstream boundary condition is more difficult, because the salinity concentration varies over a tidal cycle. This can be solved by assuming that during flood the concentration coincides with the sea concentration. During ebb, the concentration is controlled by the conditions upstream from the mouth and at the end of ebb, the salt concentration is lower than the concentration at sea. There will be a transition time for the concentration at ebb to reach the sea concentration. This transition period depends on the sea conditions and it can be determined by using the method of Thatcher and Harleman (1981).

Initial conditionals must be specified for the tidal motion, river flow and salinity. Because of the rapid convergence of the tidal hydraulics equations, it is feasible to generate initial conditions for water surface elevation and discharge. After several computational steps (e.g. 10 tidal periods), the desired start of the salinity prediction emerges (Harleman and Thatcher, 1974).

2.5 TIDAL DYNAMICS IN ESTUARIES

Many studies have been performed to describe the tidal dynamics in estuaries, ranging from mentioning and analysing a number of tidal dynamic characteristics and behaviours (e.g. Ippen, 1966; Parker, 1991; Cartwright, 1999; or Savenije, 2005) to investigating particular issues, such as tidal computations (e.g. Dronkers, 1964; or Parker *et al.*, 1999); interaction between river discharge and tides (e.g. Vongvisessomjai, 1987; or Horrevoets *et al.*, 2004); tidal wave propagation (e.g. Dronkers, 1964; Prandle and Rahman, 1980; Jay, 1991; Friedrichs and Aubrey, 1994; Lanzoni and Seminara, 1998; Godin, 1999; or Savenije and Veling, 2005) or tidal damping (Savenije, 1998, 2001), etc. In this section, firstly, we briefly introduce the general characteristics of tidal dynamics in estuaries and approaches applied to investigate the tidal dynamics. The second part of this section focuses on the tidal dynamic behaviours when propagating into an estuary. The main focus is on tidal wave celerity, tidal damping/amplification and tidal phase lag.

2.5.1 Tidal dynamics in estuaries, an introduction

Tides are the periodic upward and downward motion of the water surface (vertical tide) and the subsequent lateral flow of water (horizontal tide). Tides are generated by two main groups of forces: (i) the gravitational forces acting between the earth and the moon and between the earth and the sun; and (ii) the coriolis acceleration related to the orbital motion of the earth (Dronkers, 2005). Tides are generally classified into four types based on the lunar and solar harmonic components: semidiurnal (e.g. in Vlissingen, the Netherlands; or Immingham, UK); mixed, predominantly semidiurnal (e.g. in Soc Trang, Vietnam; or San Francisco, USA); mixed, predominantly diurnal (e.g. in Manila, Philippines; or Bangkok, Thailand); and diurnal (e.g. in Do Son, Vietnam). As the sun, the earth, and the moon move along their elliptical orbits, they continually change their positions relative to each other. As a result, the total potential defining the height of the astronomical tide changes as a function of geographic location and over time. The most prominent tidal period is the fortnightly spring–neap cycle. Due to the 28-day orbital motion of the moon, the moon-earth and sun-earth axes approximately coincide every fortnight. Thus, spring tides occur shortly after full moon and new moon, whereas neap tides occur at half-moon.

The vertical rise and fall of water surface is generally referred to as the tide and the accompanying horizontal movement is referred to as the tidal current, with the tidal flow into an estuary called the flood and the outflow of an estuary called the ebb. When a tidal wave reaches the shallower water of an estuary, it is slowed down, amplified and/or distorted due to the interaction with the estuarine topography, river discharge and due to friction. The longitudinal tidal range pattern in an estuary depends on the relative relation between convergence and friction, therefore it appears in one of three forms: amplified, damped or ideal (where the tidal range is constant - see Section 2.2.3). There are three types of tidal wave propagating into an estuary: progressive wave, standing wave and mixed wave.

- Progressive wave: The crest (i.e. high water) of the wave moves progressively inland, as does the trough (i.e. low water) of the wave. The maximum flood current is at the same time as the crest and the maximum ebb current is at the same time as the trough. The phase lag between High Water (HW) and High Water Slack (HWS) as well as between Low Water (LW) and Low Water Slack (LWS) is $\pi/2$. The wave celerity is $\sqrt{gh}$. The purely progressive wave only occurs in a frictionless channel with constant cross section and infinite length, which is not the case for real estuaries.

- Standing wave: This type of wave usually occurs in a semi-enclosed body (e.g. a bay or a river with a weir). The tidal wave is reflected at the head of the semi-enclosed body and travels back down the waterway toward the ocean. The crest and trough seem to progress at infinite speeds, whereby high water and low water occur instantaneous, with the greatest tidal range at the head of the semi-enclosed body. The tidal range decreases from the head toward the ocean, and, if the body is long enough, reaches a minimum at one location (called a node – at one-fourth of a tidal wavelength from the head) and then starts increasing again to a new maximum (called an antinode). The phase lag between HW-HWS and LW-LWS is zero. The wave celerity is infinitely large.

- Mixed wave: This type of wave occurs in an alluvial coastal plain estuary, which has a channel with non-constant cross section (e.g. gradually exponential shape). The longitudinal tidal range pattern depends on the relative relation between convergence and friction; therefore, it appears in one of the three forms mentioned above. The phase lag is between 0 and $\pi/2$. The wave celerity differs from the progressive wave celerity, depending on the damping or amplifying characteristics.

Tidal dynamics in estuaries can be investigated and analysed numerically or analytically. Equations 2.48 and 2.49, in fact, are the Saint-Venant's equations. These two equations can be used to determine the hydrodynamic regimes of estuaries. Initially, analytical solutions were the only option to turn these non-linear differential equations into practical applications. An example can be seen in Ippen (1966). Later, despite the quick development of computers and computer programming, a number of authors have derived analytical solutions (Jay, 1991; Friedrichs and Aubrey, 1994; Lanzoni and Seminara, 1998; Godin, 1999; Savenije, 2001; Prandle, 2003; Horrevoets *et al.*, 2004; and Savenije and Veling, 2005). The advantage of these analytical equations is that they provide insight into the functioning of the hydraulics and can be used to assess the outcomes of numerical models.

Generally, with assistance of computers, numerical models can provide good solutions on the hydrodynamic regimes of a considered study area. However, the solutions obtained depend on several factors: (i) accuracy and stability of the numerical schemes; (ii) correctness of the topography; (iii) accuracy of the boundary conditions; and (iv) good set of prototype data and observations for calibration and validation.

2.5.2 Tidal wave characteristics

Analytical approaches can provide analytical solutions of the tidal dynamic equations by solving the conservation of mass and conservation of momentum equations. Most authors obtained analytical solutions using perturbation analysis, where the scaled equations are simplified by neglecting higher order terms. A number of authors (e.g. Dronkers, 1964; Hunt, 1964; Prandle and Rahman, 1980; Parker, 1984; or Jay, 1991) assumed that: (i) The local acceleration contributes to momentum at first order; and (ii) discharge gradients due to velocity variation contribute to mass at first order. These terms are sometime important, especially for strongly convergent estuaries such as the Delaware, Thames, Incomati or Pungue. Friedrichs and Aubrey (1994) derived second-order solutions and they agreed well with observations of the Thames, Tamar and Delaware. Savenije (2005) used a simple harmonic solution without simplifying the equations (i.e. complete non-linearized Saint-Venant equations) and obtained good agreements with observations in the Scheldt and Incomati. However, they are implicit analytical equations. Subsequently, Savenije *et al.* (2007) improved the set of these equations into a new set of explicit equations and gained good agreements with observations and numerical model results of the Elbe and the Scheldt.

As we have seen earlier, the characteristics of the tidal wave propagating into an estuary can be described through three main factors: tidal wave celerity, phase lag and tidal range variability (i.e. tidal damping/amplification). Therefore, we proceed to briefly introduce the analytical solutions of Savenije *et al.* (2007) to investigate these factors.

For given topography, friction and tidal amplitude at the downstream boundary, the velocity amplitude, the wave celerity, the tidal damping and the phase lag can be computed. The set of implicit equation includes (Savenije, 2005):

The phase lag equation: $$\tan\varepsilon = \frac{\omega a}{c} \Big/ \left(1 - \frac{a}{\eta}\frac{\mathrm{d}\eta}{\mathrm{d}x}\right) \tag{2.51}$$

The scaling equation: $$r_S \frac{\eta}{\overline{h}} = \frac{\upsilon}{c}\frac{1}{\sin\varepsilon} \tag{2.52}$$

The damping equation: $$\frac{1}{\eta}\frac{\mathrm{d}\eta}{\mathrm{d}x}\left(\frac{1+\alpha}{\alpha}\right) = \frac{1}{a} - f\frac{\upsilon\sin\varepsilon}{\overline{h}c} \tag{2.53}$$

The celerity equation: $$c^2 = \frac{1}{r_S} g\overline{h} \Big/ \left[1 - \frac{\sin\varepsilon\cos\varepsilon}{(1+\alpha)}\left(\frac{c}{\omega a} - f\frac{\upsilon\sin\varepsilon}{\omega\overline{h}}\right)\right] = \frac{c_0^2}{1-D} \tag{2.54}$$

where ε (T) and ω (T^{-1}) are the phase lag between HW-HWS and angular velocity. η (L) and $\overline{h}$ (L) are tidal amplitude and tidal average depth. υ (LT^{-1}) is tidal amplitude. c (LT^{-1}) and $c_0 = \sqrt{g\overline{h}/r_S}$ (LT^{-1}) are the actual and classical tidal wave celerities, respectively. f (-) is the friction factor, D (-) is the damping term and α (-) is the tidal Froude number: $\alpha = (c\upsilon\sin\varepsilon)/(g\eta)$.

These four equations (i.e. Eqs. 2.51 - 2.54) can be scaled making use of the scales used by Toffolon *et al.* (2006). The four equations then become:

The phase lag equation: $$\tan\varepsilon = \frac{\lambda}{\gamma - \delta} \tag{2.55}$$

The scaling equation: $$\mu = \frac{\sin\varepsilon}{\lambda} = \frac{\cos\varepsilon}{\gamma - \delta} \tag{2.56}$$

The damping equation: $$\delta = \frac{\mu^2}{\mu^2 + 1}\left(\gamma - \chi\mu^2\lambda^2\right) \tag{2.57}$$

The celerity equation: $$\lambda^2 = 1 - D = 1 - \delta\frac{\cos\varepsilon}{\mu} = 1 - \delta(\gamma - \delta) \tag{2.58}$$

where γ (-) is the estuary shape number: $\gamma = c_0/\omega a$; χ (-) is the friction number: $\chi = r_S f\left(c_0/\omega\overline{h}\right)\zeta$ (where $\zeta = \eta/\overline{h}$). μ (-) and λ (-) are the velocity and celerity numbers, respectively. δ (-) is the tidal damping number : $\delta = \frac{1}{\eta}\frac{\mathrm{d}\eta}{\mathrm{d}x}\frac{c_0}{\omega}$.

This is an implicit set of four equations computing the velocity amplitude, the wave celerity, the tidal damping and the phase lag. The set can be solved iteratively, but can also be solved by an explicit analytical solution. Savenije *et al.* (2007) derived two different families of solutions: (i) family for mixed tidal waves; and (ii) family for standing waves. Figure 2.4 presents the solution of the set of four equations.

It appears in Fig. 2.4 that there are two distinct types of estuaries. For the strongly convergent estuaries (i.e. large values of γ), they no longer depend on the friction number χ, implying that they correspond to the class of estuaries with a standing tidal wave. For the weakly convergent estuaries (i.e. small values of γ), they strongly depend on χ and they correspond to the class of estuaries with a mixed tidal wave. Progressive waves correspond to the case where $\gamma=0$ and $\chi=0$. Figure 2.4 also presents lines indicating the sub-set of ideal estuaries, where there is no damping (because there is an equilibrium between friction and convergence), and where $c = c_0\,(\lambda = 1)$.

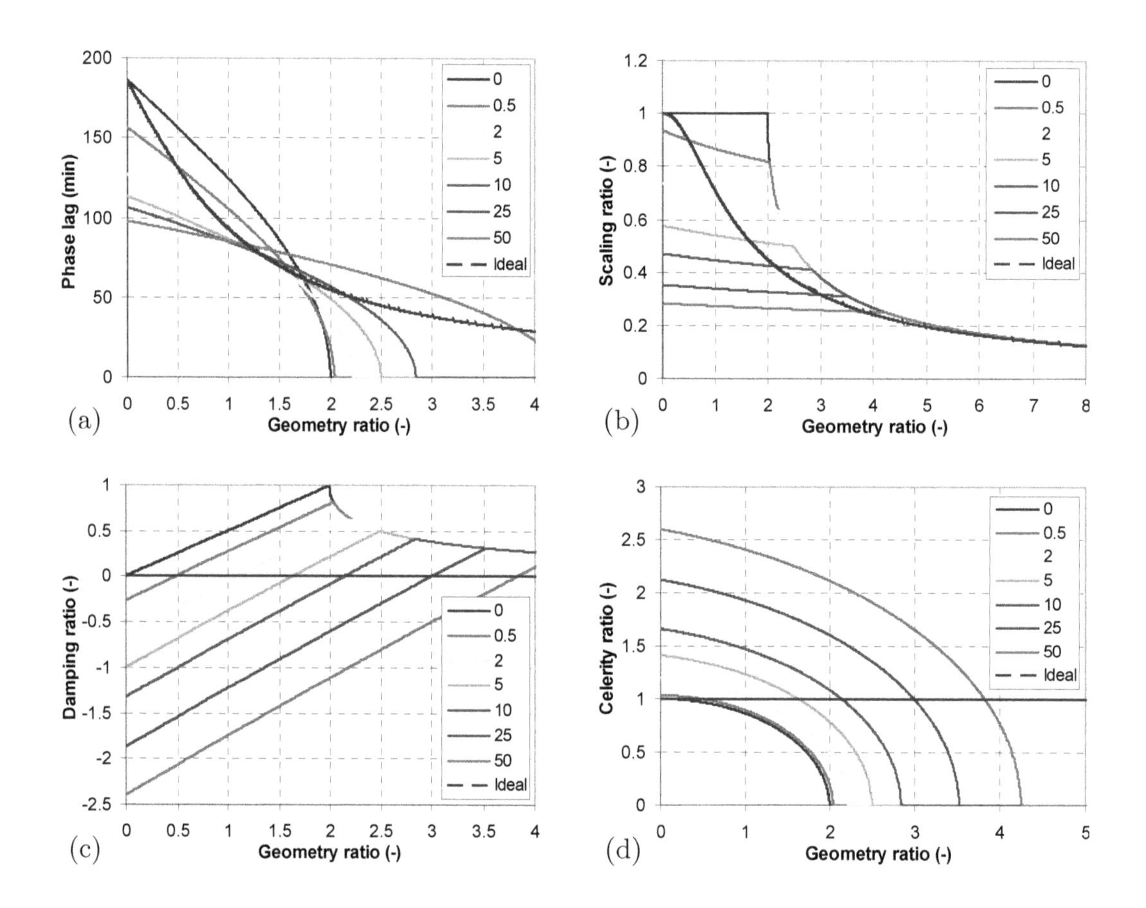

Figure 2.4 Relationship between: (a) phase lag ε; (b) the velocity number μ; (c) the tidal damping number δ; and (d) the wave celerity number λ and the estuary shape number γ for different values of the friction number χ.

It is concluded in Savenije *et al.* (2007) that the analytical solutions offer an opportunity to look into the functioning of the Saint-Venant's equations without the need for a complex hydrodynamic model. It is recommended that the classification of estuaries should be based on two parameters, i.e. the estuary shape γ and the friction scale χ.

2.6 CONCLUSIONS

This chapter provides an overview on classification of estuaries, mixing, salinity intrusion mechanisms and tidal wave characteristics in estuaries. It can be seen that due to their abundant and diverse characteristics, there are many ways to classify estuaries. However, it is clear that the two dominant drivers of an estuary are its tide and river discharge. They are in fact the main two factors causing mixing and driving salt intrusion in estuaries. Tidal pumping caused by residual ebb-flood channel circulation in multi-channel estuaries, which is the main tide-driven mixing mechanism, is expected to play a major role in producing the longitudinal salt dispersion. However, to date no effort has been made to quantify this mechanism in terms of a longitudinal salt dispersion. In addition, steady-state salt balance models and predictive salt balance models have only been developed for single-branch estuaries. And finally, there are not many analyses paying attention to tidal wave characteristics in multi-channel estuaries, with respect to tidal wave celerity, phase lag and tidal range variability.

Chapter 3

SALT INTRUSION IN MULTI-CHANNEL ESTUARIES

Abstract

There is a well-tested theory for the computation of salt intrusion in alluvial estuaries that is fully analytical and predictive. The theory uses analytical equations to predict the mixing behaviour of the estuary based on measurable quantities, such as channel topography, river discharge and tidal characteristics. It applies to single-channel topographies and estuaries that demonstrate moderate tidal damping. The Mekong Delta is a multi-channel estuary where the tide is damped due to a relatively strong river discharge (in the order of 2,000 m^3/s), even during the dry season. As a result, the Mekong is a strongly riverine estuary. This chapter aims to test if the theory can be applied to such a riverine multi-channel estuary, and to see if possible adjustments or generalisations need to be made. The chapter presents salt intrusion measurements that were done by moving boat in 2005, to which the salt intrusion model has been calibrated. The theory has been expanded to cater for tidal damping. Subsequently the model has been validated with observations made at fixed locations over the years 1998 and 1999. Finally, it has been tested whether the Mekong calibration fits the overall predictive equations derived in other estuaries. The test has been successful and led to a slight adjustment of the predictive equation to cater for estuaries that experience a sloping bottom.

3.1 INTRODUCTION

The recent book on salt intrusion and tides in alluvial estuaries (Savenije, 2005) presents a comprehensive theory for the modelling of steady state and unsteady state salt intrusion in alluvial estuaries. It is based on the analysis of some 17 estuaries world-wide on the basis of which a general predictive theory has been developed that

---

This chapter was published as:

Nguyen, A.D., and Savenije, H.H.G., 2006. Salt intrusion in multi-channel estuaries: A case study in the Mekong Delta, Vietnam. Hydrology and Earth System Sciences 10: 743-754.

is claimed to be predictive, meaning that it can be applied to predict the salinity distribution in any alluvial estuary provided some basic information on topography, tide and river discharge is known (Savenije, 1986, 1989, 1993c). Limitations of the theory are that it has been derived for single channel estuaries and for estuaries where the tide experiences only modest damping or amplification.

The Mekong Delta is an alluvial estuary that consists of eight branches and that transports a large amount of fresh water to the sea, even during the dry season (in the order of 2,000 m^3/s). As a result, the tide is strongly damped and the branches of the delta are rather prolonged. This gives the Mekong estuary a clearly riverine character, putting it at the fringe of applicability of this theory.

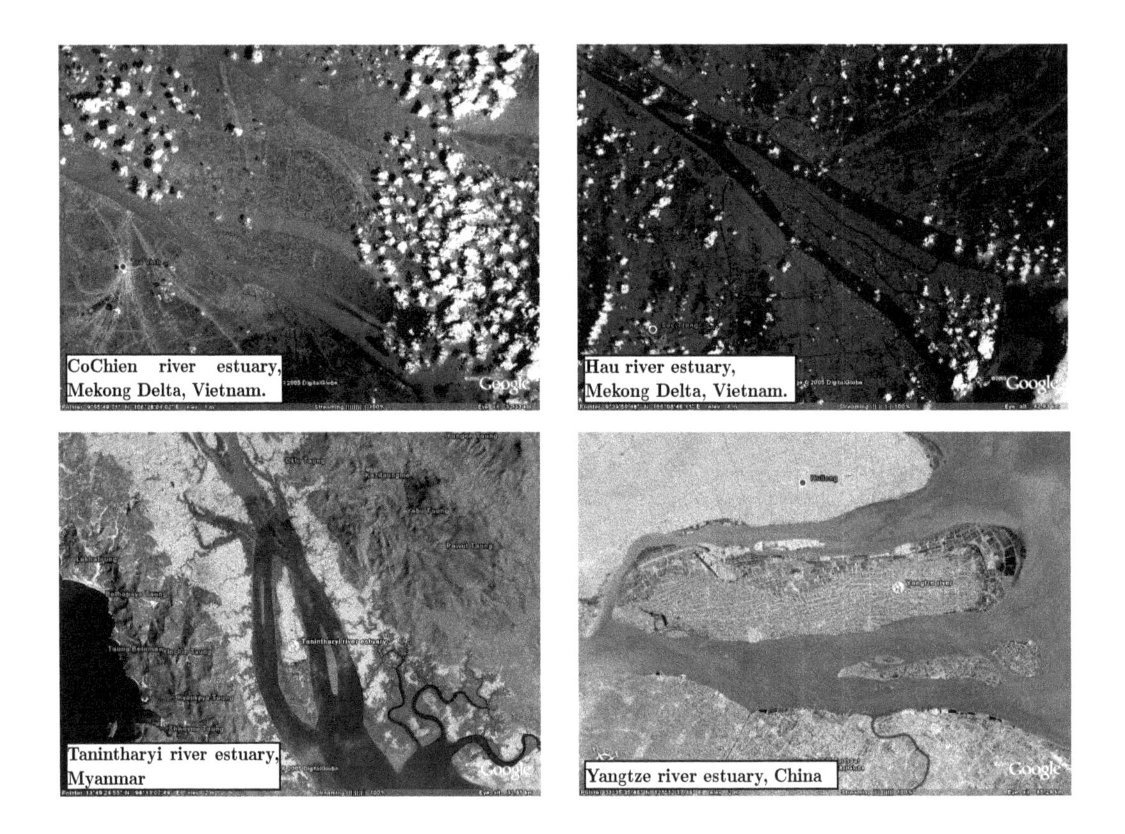

Figure 3.1 Some examples of branched estuaries (source: Google Earth – not to scale)

Looking at other branched alluvial estuaries in the world, it appears that they have similar characteristics: (i) if an estuary is divided into two branches by an island, the branches are of almost equal length; and (ii) there are no large differences between the branches in area, width and depth. Examples of branched estuaries are the Loire estuary in France, the Tanintharyi estuary in Myanmar, the Hau and Co Chien branches of the Mekong Delta in Vietnam, the Dhamra estuary in India and the Yangtze river estuary in China (see Fig. 3.1).

The question now is: does the method of Savenije (2005) also apply to these estuaries, or is there a need to extend the theory so that it can be applied to branched estuaries as well. In this chapter, we shall demonstrate that the method is indeed applicable to branched estuaries but that a certain procedure needs to be followed. We also present some refinements to the method for estuaries that experience strong tidal damping or that have a prominent bottom slope.

3.2 SUMMARY OF THE METHOD

Savenije (2005, p. 144) demonstrated that the steady state salt balance equations for High Water Slack (HWS), Low Water Slack (LWS) and Tidal Average (TA) situation can be written as:

$$S_i - S_f = c_i \frac{\mathrm{d}S_i}{\mathrm{d}x} \tag{3.1}$$

where i indicates the three different states: HWS, LWS and TA. S_i (-) is the steady state salinity, S_f (-) is the fresh water salinity. The coefficient c_i is an x-dependent coefficient defined as:

$$c_i = \frac{A}{Q_f} D_i \tag{3.2}$$

in which D_i (L^2T^{-1}) is the dispersion coefficient for each state i, Q_f (L^3T^{-1}) is the river discharge, which is negative since the positive x-axis points upstream, and A (L^2) is the tidal averaged cross-sectional area.

The relationship between the salinity and the dispersion coefficient, based on previous work by Van den Burgh (1972), is defined by:

$$\frac{\mathrm{d}D_i}{\mathrm{d}x} = K \frac{Q_f}{A} \tag{3.3}$$

where K (-) is van den Burgh's coefficient, which has a value between 0 and 1. This equation can be integrated for an estuary with an exponentially varying cross-section (see Eq. 3.11) to yield the expression for the dispersion along the estuary:

$$\frac{D_i}{D_{0i}} = 1 - \beta_i \left(\exp\left(\frac{x}{a} \right) - 1 \right) \tag{3.4}$$

with:

$$\beta_i = -\frac{K a Q_f}{D_{0i} A_0} = \frac{K a}{\alpha_{0i} A_0} \tag{3.5}$$

and:

$$\alpha_{0i} = -\frac{D_{0i}}{Q_f} \tag{3.6}$$

in which D_{0i} (L^2T^{-1}) is the boundary condition at the river mouth (x=0) for HWS, LWS or TA conditions, A_0 (L^2) is the tidal average cross-sectional area at the estuary mouth and a (L) is the convergence length of the cross-sectional area (see Eq. 3.11). α_{0i} (L^{-1}) is the mixing coefficient at the estuary mouth. The values of K and α_{0i} can be obtained through calibration against measured longitudinal salinity distributions at HWS or LWS.

The longitudinal variation of the salinity can be computed through combination of Eqs. 3.1, 3.2 and 3.3:

$$\frac{S_i - S_f}{S_{0i} - S_f} = \left(\frac{D_i}{D_{0i}}\right)^{\frac{1}{K}} \tag{3.7}$$

where S_{0i} (-) is the boundary salinity at the estuary mouth (for HWS, LWS and TA conditions). The salt intrusion curve derived for the TA situation, which represents the TA longitudinal variation of the salinity, can be used for LWS and HWS as well, by shifting the curve upstream or downstream over half the tidal excursion E.

The salt intrusion length L_i can be obtained at the point that $S_i = S_f$, then according to Eq. 3.7, D_i equals to zero. With $D_i = 0$, Eq. 3.4 can be elaborated to yield an expression for the intrusion length:

$$L_i = a \ln\left(\frac{1}{\beta_i} + 1\right) \tag{3.8}$$

where L_i (L) is the salt intrusion length at HWS, LWS or TA.

Because the method has been applied in 17 different estuaries all over the world, particularly for the HWS situation, it was possible to derive two predictive equations for K and D_0^{HWS} (Savenije, 1993c). These relations were generalised and improved by Savenije (2005, pp. 166-169) into:

$$\frac{D_0^{HWS}}{\upsilon E} = 1400 \frac{\bar{h}}{a} \sqrt{N_R} \tag{3.9}$$

and:

$$K = 0.2 \times 10^{-3} \left(\frac{E}{H}\right)^{0.65} \left(\frac{E}{C^2}\right)^{0.39} (1 - \delta b)^{-2.0} \left(\frac{b}{a}\right)^{0.58} \left(\frac{Ea}{A_0}\right)^{0.14} \tag{3.10}$$

in which A_0 (L^2) is the cross-sectional area at the estuary mouth, b (L) is the width convergence length; C ($L^{0.5}T^{-1}$) is the Chezy coefficient, δ (L^{-1}) is the damping rate of tidal range; E (L) is the tidal excursion, which is the distance that a water particle travels between LWS and HWS, obtained from the observed salinity distributions; υ (LT^{-1}) is the tidal velocity amplitude; H (L) is the tidal range, $\bar{h}$ (L) is the constant tidal average depth along the estuary, N_R (-) is the Estuarine Richardson number given by Eq. 2.2, and T (T) is the tidal period.

In ideal estuaries, the tidal range, the tidal velocity amplitude, the tidal excursion and the depth are constant along the estuary, while the convergent lengths of the width and the cross-sectional area are equal: $b=a$. In estuaries where there is a certain degree of damping or amplification, the ratio of H/E is still constant, but values of E and υ vary along the estuary. Finally if $a \neq b$ there is a bottom slope and the depth is not constant. For the case of the Mekong estuary branches, the latter applies, and subsequently special procedures need to be developed.

The assumption made in Eq. 3.1 to arrive at the steady state equation for conservation of mass, requires that in the estuary an equilibrium condition is reached between, on the one hand, advective salt transport through the downstream flushing of salt by the fresh water discharge, and on the other hand, the full range of mixing processes. To investigate how quickly an estuary system adjusts to a new situation, Savenije (2005, p. 152) proposed an expression derived for the system response time as a function of the steady state salinity distribution (see Chapter 2, Section 2.4.2).

3.3 THE MEKONG DELTA IN VIETNAM

3.3.1 Overview

In Chapter 1, the Mekong Delta has already been introduced. Here the relevant characteristics of the delta are briefly summarized. The Mekong river when it enters Vietnam splits into two branches, the Bassac (known as the Hau river in Vietnam) and the Mekong (known as the Tien river in Vietnam). The two branches form the Mekong Delta.

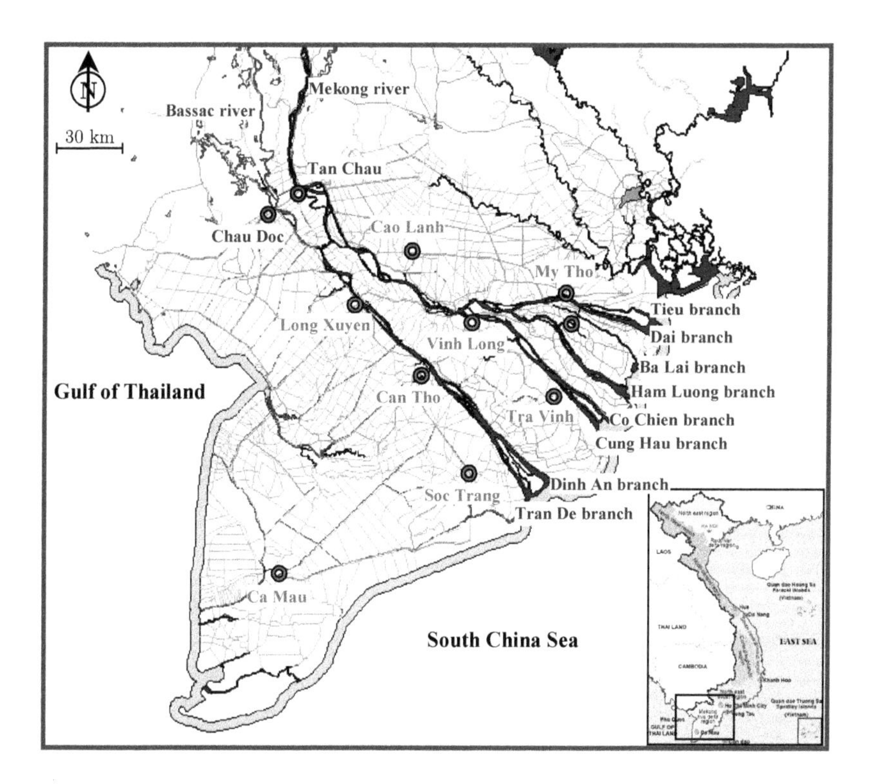

Figure 3.2 Branches of the Mekong Delta

The Hau river is the most southern branch of the river system. When the Hau approaches the sea, it splits into two branches: Tran De and Dinh An. The Tien river is the northern branch of the river system. At Vinh Long, the Tien separates into two river branches: Co Chien and My Tho. At a distance of 30 km from the South China Sea, the Co Chien river again splits into two estuary branches, Co Chien and Cung Hau (see Fig. 3.2). Although there are more estuary branches: Tieu, Dai, Ba Lai and

Ham Luong, this chapter concentrates on the four estuary branches: Tran De, Dinh An, Cung Hau and Co Chien.

Tides in the South China Sea have a mixed diurnal and semi-diurnal character. The amplitude can be up to 3 m. There are generally two troughs and two peaks during a day, but their relative height varies over a fortnight. When the first trough decreases from day to day, the other trough increases, and vice versa (See Fig. 3.3).

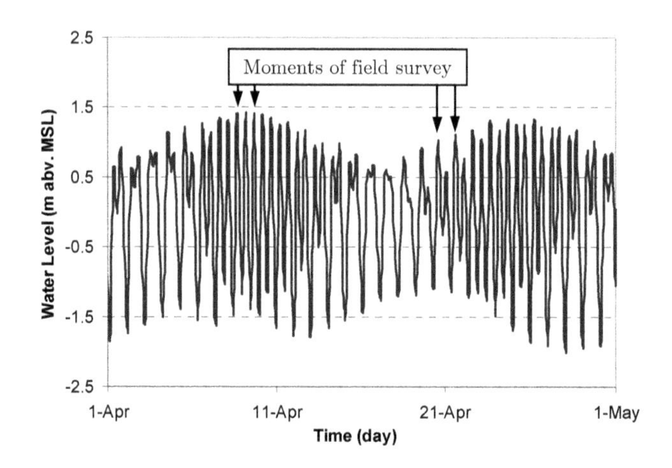

Figure 3.3 Water level at the My Thanh station located near the mouth of the Tran De estuary branch (Period from 1 April 2005 to 1 May 2005)

As a typical delta, the Mekong Delta is affected by both river floods and tides. In the past (before 1980), every year during the dry season, agricultural areas in the Mekong Delta were affected by salinity, amounting to 1.7–2.1 million ha out of 3.5 million ha. In the 1980's and 1990's, a number of salinity control projects were implemented, leading to closure dams and sluice gates in the navigation canals connecting the branches of the delta. Nowadays, salinity affects only 0.8 million ha every year. However, the fresh water intakes along the estuary branches are usually affected by high salinity (Nguyen and Nguyen, 1999). Every year, these intakes have to be closed for considerable periods (varying from several weeks to one or two months) to prevent salt intrusion.

3.3.2 The shape of the Mekong Delta branches

The Dinh An, Tran De, Co Chien and Cung Hau are four branches of a branched estuary system. We shall investigate if it is possible to combine paired branches into a single branch, which would not only simplify the computation, but could also enhance the overall performance of the salt intrusion model. Moreover since we expect that the estuary system functions as an entity, it could very well be that combining paired branches into a single branch is more in agreement with the physical laws that guide the formation of ebb and flood channels than a separate treatment.

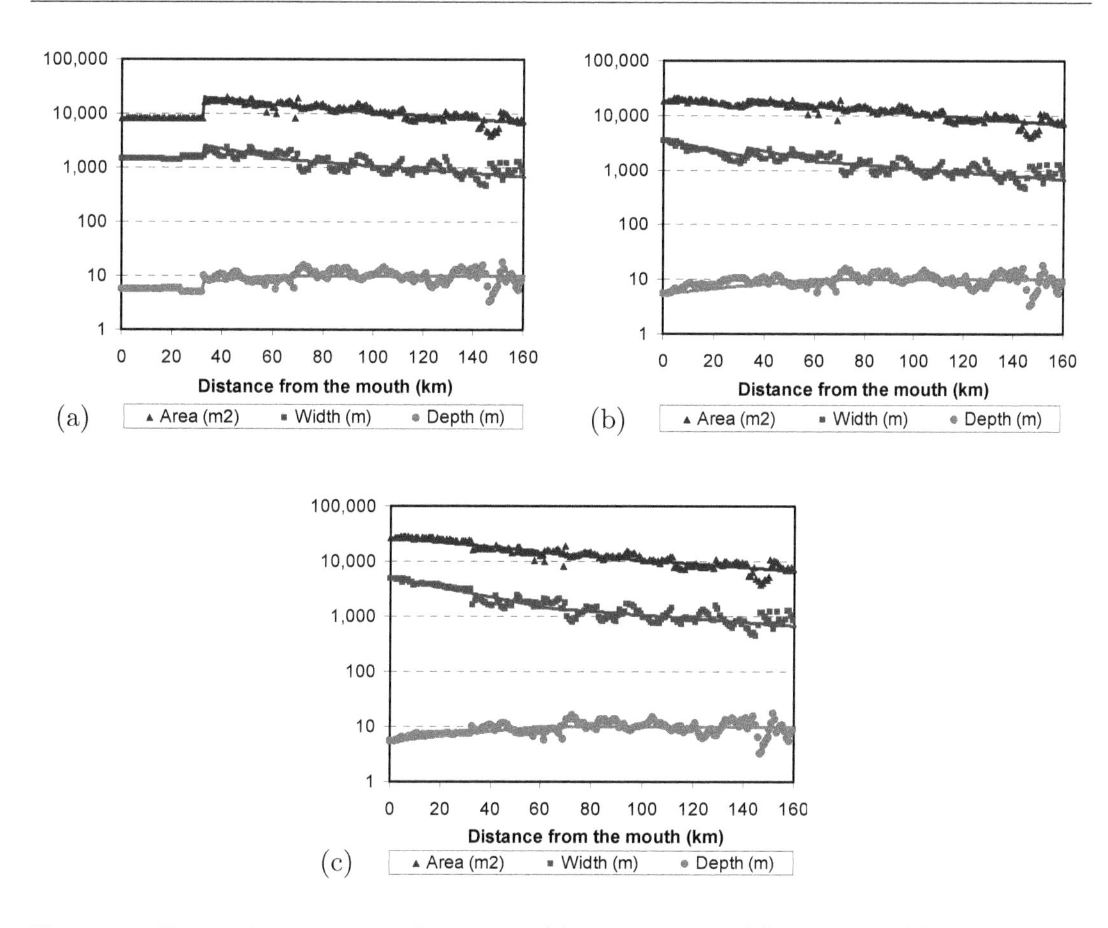

Figure 3.4 Shape of the individual branches (a) Tran De and (b) Dinh An. (c) Shape of the Hau (combination of Tran De and Dinh An) estuary, showing the measured area (triangles), measured width (squares) and measured depth (circles).

Table 3.1 Estuarine characteristics of 4 branches in the Mekong Delta

River	Estuary	A_0 (m^2)	B_0 (m)	$\bar{h}$ (m)	a (km)	b (km)	d (km)	a_1 (km)	b_1 (km)
Hau river	Dinh An branch	18,400	3,400	7.6	100	47	89	-	-
	Tran De branch	8,200	1,500	5.5	800	800	∞	-	-
	Combined estuary	26,600	4,900	7.5	105	51	99	140	140
Co Chien river	Co Chien branch	11,100	1,600	7.6	93	71	300	-	-
	Cung Hau branch	13,200	2,500	5.7	45	40	360	-	-
	Combined estuary	24,300	4,100	6.6	69	54	250	-	-

Note: The values for the width, depth and cross-sectional area were measured at Mean Sea Level (MSL). a_1 and b_1 are the area and width convergence lengths after the inflection point, respectively.

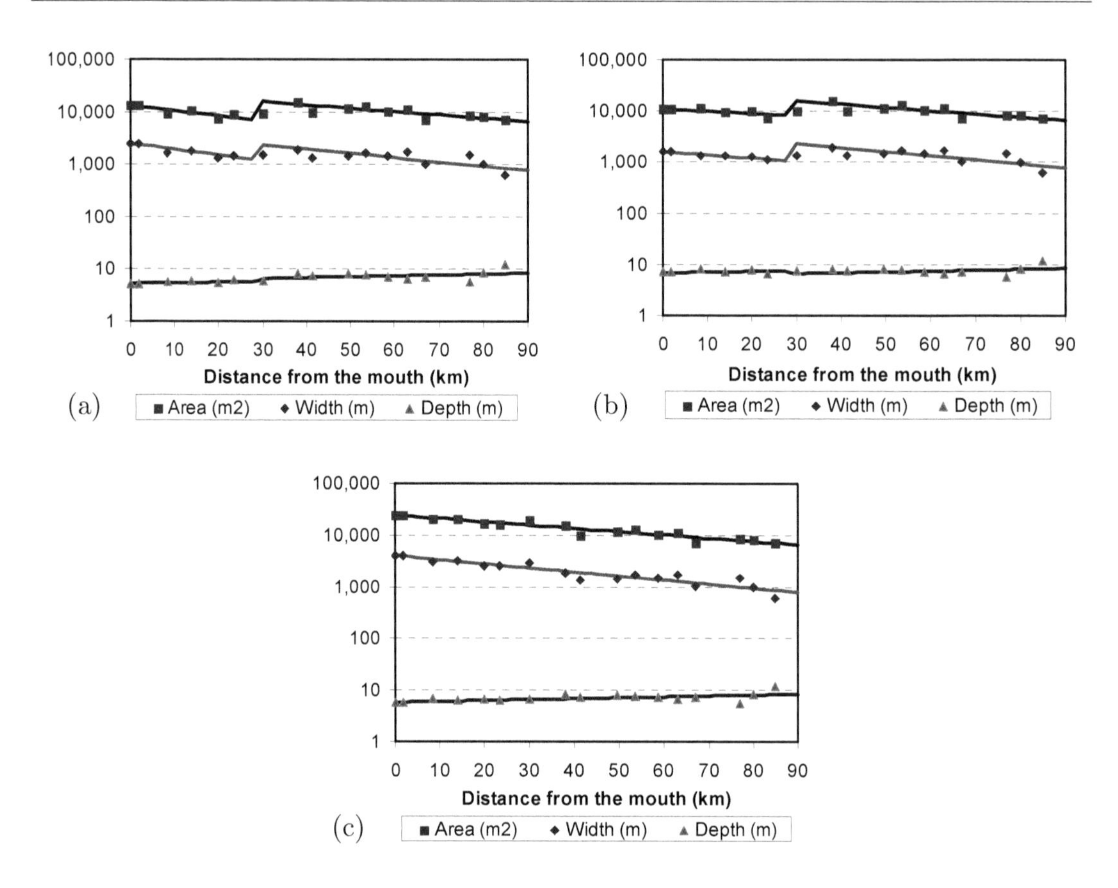

Figure 3.5 Shape of the individual branches: (a) Cung Hau and (b) Co Chien. (c) Shape of the Co Chien - Cung Hau (combination of Co Chien and Cung Hau) estuary, showing the measured area (squares), measured width (diamonds) and measured depth (triangles).

The estuary shape of the Tran De - Dinh An and the Co Chien - Cung Hau can be described by the exponential functions:

$$A = A_0 \exp\left(-\frac{x}{a}\right) \tag{3.11}$$

$$B = B_0 \exp\left(-\frac{x}{b}\right) \tag{3.12}$$

$$h = h_0 \exp\left(\frac{x}{d}\right) \tag{3.13}$$

where A (L^2), B (L) and h (L) are the cross-section area, width and depth at the location x (km) from the mouth, respectively. A_0 (L^2), B_0 (L) and h_0 (L) are the area, width and depth at the mouth. Finally, a (L), b (L) and d (L) are the area, width and depth convergence length, respectively. It follows that $d=ab/(a\text{-}b)$. The cross-section area and width are obtained from observations, defined at the tidal averaged water level (this level is close to the mean sea level). The convergence length, which is the length scale of the exponential function, is obtained by calibration of Eqs. 3.11, 3.12 and 3.13 against measured data. It can be seen very clearly in Fig. 3.4, Fig. 3.5 and

Table 3.1 that the paired branches indeed behave as a single estuary with a regular topography according to Eqs. 3.11, 3.12 and 3.13. The combined Hau estuary has an inflection point located 57 km from the Dinh An mouth and upstream from the inflection point, the shape of the estuary also well corresponds with exponential functions, while the depth is constant.

3.4 DATA SETS OF THE MEKONG DELTA

Salinity data of the Mekong Delta used in this chapter consist of two sets: (i) the first set is from field measurements carried out by the author during the dry season of 2005; and (ii) the second set is from data of a network of fixed stations along the estuaries.

The author carried out field measurements on the Hau river in the dry season of 2005, using the "moving boat" method described by Savenije (1989, 1992a). The method applied to the Mekong Delta estuaries can be summarized as follows: To obtain the salt intrusion curve at HWS or LWS one has to travel at the same speed as the tidal wave. Before the start of the measurement, one identifies clearly recognizable points along the estuary at about 3-km interval (inlets, piers, churches, temples, gauges, etc.). Additionally a GPS device is used to obtain the exact co-ordinates of the locations. The measurement starts at the mouth in mid-stream. While anchored one waits until the moment of slack occurs. The first observation is done at slack tide, measuring at different points over the depth. Travelling at a velocity of 25-30 km/h, one can arrive at the next location just a few minutes before slack. One then stops the boat and waits until the slack moment occurs (either by anchoring or by closely watching the shores or the GPS device as one drifts). At slack, one does a quick measurement over the full depth at one-meter intervals starting from the bottom with a conductivity meter with a 10-m cable. Measurements are done in mid-current.

The first and second survey were carried out at the moment of LWS and HWS on 8 and 9 April during spring tide in the Tran De and Dinh An branches. The third and fourth survey in the Hau river were conducted on 21 May in Tran De and on 22 May in Dinh An. The field measurements in the Co Chien and Cung Hau estuary branches were carried out on 21 and 22 April at the moments of LWS and HWS during spring tide. The second data set is obtained from the network of fixed stations near intakes and quays, which measure salinity values during the dry season at hourly intervals.

During the moving boat measurements, we managed to measure vertical salinity distribution at several points in the Dinh An, Tran De, Co Chien and Cung Hau branches. It appeared that these branches are partially-mixed and well-mixed estuaries corresponding to the estuarine stratification classification (see Section 2.2.4, Chapter 2).

River discharge and tidal data during the 2005's field measurements provided by the Vietnamese National Hydrometeorology Services (VNHS) are summarized in Table 3.2. One can see that within two successive days river discharge and tidal range variations are small, and therefore salinity variations between successive HWS and LWS situations are supposed to be small as well.

Table 3.2 River discharge and tidal range data in the Mekong estuaries

	Estuary name	Date	Tidal range (m)	River discharge (m^3/s) (*)
Mekong Delta	Hau	08- April - 2005	2.89	2,396
		09- April - 2005	2.90	2,345
		21 – May - 2005	2.62	2,021
		22 – May - 2005	2.75	2,167
	Co Chien – Cung Hau	21- April - 2005	2.07	2,227
		22- April - 2005	2.14	2,017

(*) Total discharge of both the Tien (Tan Chau station) and the Hau river (Chau Doc station), about 30 km upstream of the Vam Nao connection.

3.5 Salinity computation for the Mekong Delta branches

In this section, two approaches are used to compute the salinity distribution in the Mekong Delta estuaries. The first approach considers every single branch as an individual estuary where the method of Savenije (2005) is used to determine the longitudinal salinity distribution.

Because of the similarities between the estuary branches, the second approach considers the combination of paired estuary branches as a single estuary. Here, the method is applied under some modifications. Compared to the first approach, the second approach offers better results, which is subsequently validated against historic observations at fixed locations. Finally, based on the calibration and validation results of the second approach, a predictive model for the Mekong Delta estuaries is proposed.

3.5.1 Approach 1: Analysis of individual branches

The first approach considers every single branch as an individual estuary (i.e. Tran De, Dinh An, Cung Hau, Co Chien). The channel topography of each estuary is shown in Table 3.1. Using Eqs. 3.4 and 3.7, we can compute the longitudinal variation of the salinity in every individual estuary. The calibration results, based on measurement data of 8, 9, 21, 22 April and 21, 22 May 2005, are presented in Fig. 3.6.

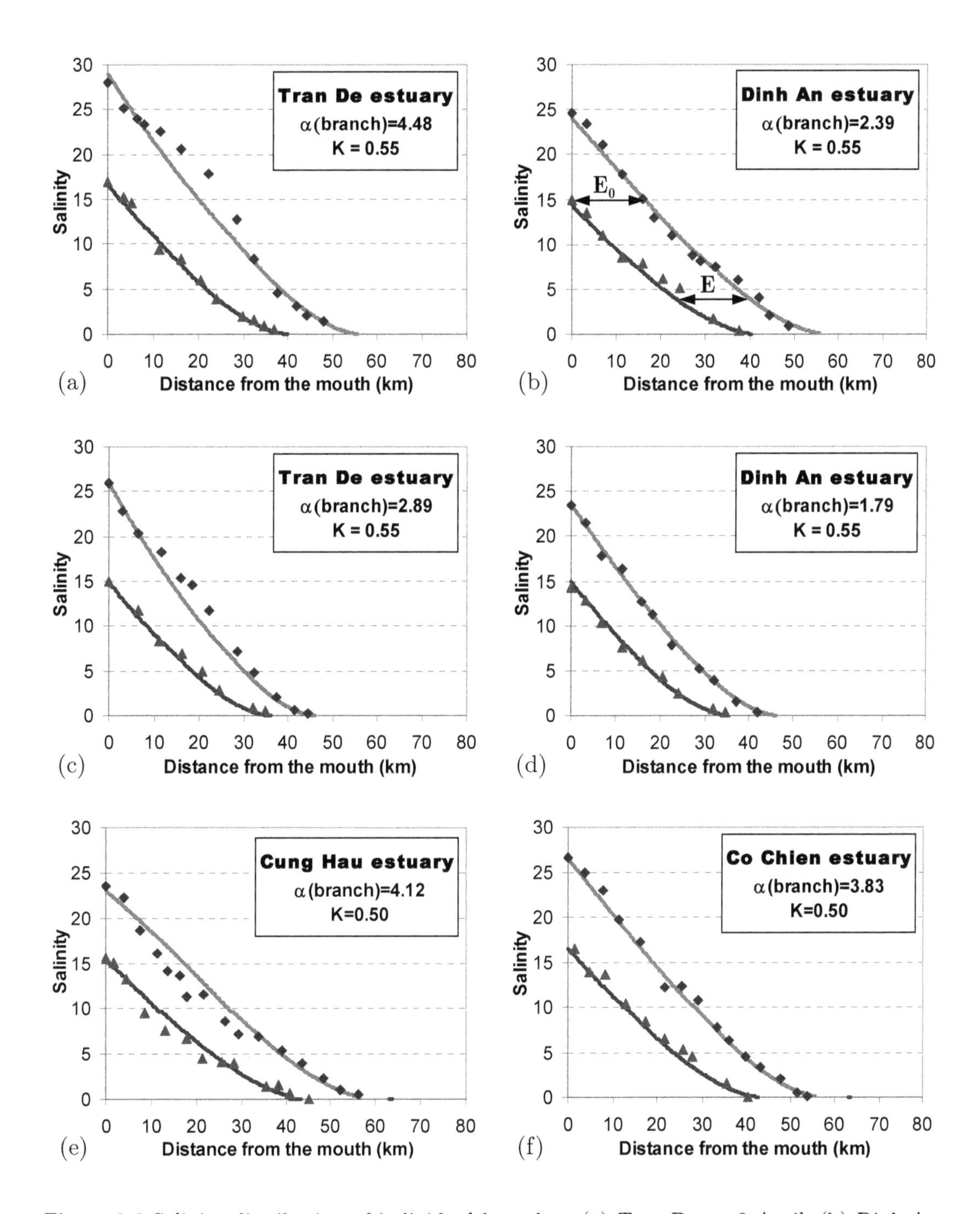

Figure 3.6 Salinity distribution of individual branches: (a) Tran De on 8 April, (b) Dinh An on 9 April, (c) Tran De on 21 May, (d) Dinh An on 22 May, (e) Cung Hau on 21 April and (f) Co Chien on 22 April 2005, showing values of measured salinity at HWS (diamonds), and LWS (triangles), and calibrated salinity curves at HWS (upper curves) and LWS (lower curves).

One thing we can conclude from the measurement data is that the tidal excursion, which is the distance that a water particle travels between LWS and HWS (and hence the horizontal distance between the drawn curves), is damped along the estuary branches. This is due to the considerable discharge of the Mekong even during the dry season. The tidal excursion in the Hau and the Co Chien - Cung Hau branches can be described by an exponential function, i.e. $E = E_0 \exp(-x/e)$ where E (L) is the tidal excursion at location x (km) from the mouth, E_0 (L) is the tidal excursion at the mouth and e (L) is the damping length. E_0 is determined by the horizontal distance between the LWS salinity curve and the HWS salinity curve at the mouth (See an example in Fig. 3.6b). E_0 and e are determined by fitting against the longitudinal salinity distributions.

We can see from Fig. 3.6 that the method can be used to describe the salinity distribution in the individual branches of the Hau and the Co Chien - Cung Hau. The measured data in the Dinh An estuary branch can be considered to be the best, thanks to the navigation buoys that gave the author a clear view of the occurrence of HWS and LWS.

3.5.2 Approach 2: Combination of two branches into a single estuary

The second approach considers the combination of two paired branches as a single estuary branch. The combination of the Dinh An and Tran De branches is named "Hau estuary" and the combination of the Co Chien and Cung Hau branches is named "Co Chien - Cung Hau estuary". The channel topography of the combined estuaries is shown in Table 3.1 and Figs. 3.4c and 3.5c. The salinity of one combined estuary is taken as a weighted mean between the cross-sectional areas of the branches. Using Eqs. 3.4 and 3.7, we can compute the longitudinal variation of the salinity in every combined estuary. The calibration results for the combined estuaries are presented in Fig. 3.7.

One can see that the second approach produces very good results. The overall performance is better than that of the first approach. This underlines the assumption that the estuary system functions as an entity.

We can observe that the mixing coefficients of the individual branches in Fig. 3.6 are always higher than the values of the combined estuaries in Fig. 3.7. This is understandable since the river discharge is split over the two branches, while the tidal range remains the same. As a result, α_0 is larger in the individual branches than in the combined branches.

3.5.3 Validation of Approach 2

In the Mekong Delta, there is a network of fixed stations near intakes and quays, which measure salinity values during the dry season at hourly intervals. Unfortunately, this information is not ideal since it does not permit the direct derivation of the longitudinal distribution of the maximum and minimum salinity (i.e. salinity at LWS and HWS), partly because they are often not located near the main current (sometimes they are located within a canal opening or intake), and partly because of the timing. However, we shall use these maximum and minimum daily values as indicators for the HWS and LWS salinity, to validate the model.

The validation results are presented in Fig. 3.8 (for the combined Hau estuary) and Fig. 3.9 (for the combined Co Chien - Cung Hau estuary). Generally, the model performs reasonably well, especially at HWS. At several stations, e.g. Tra Kha (4 km from the Dinh An mouth) and Hung My (12.7 km from the Co Chien mouth), the observed salinity values are too small and inaccurate. This is to be expected since these stations are located in the mouth of a canal or close to the riverbanks.

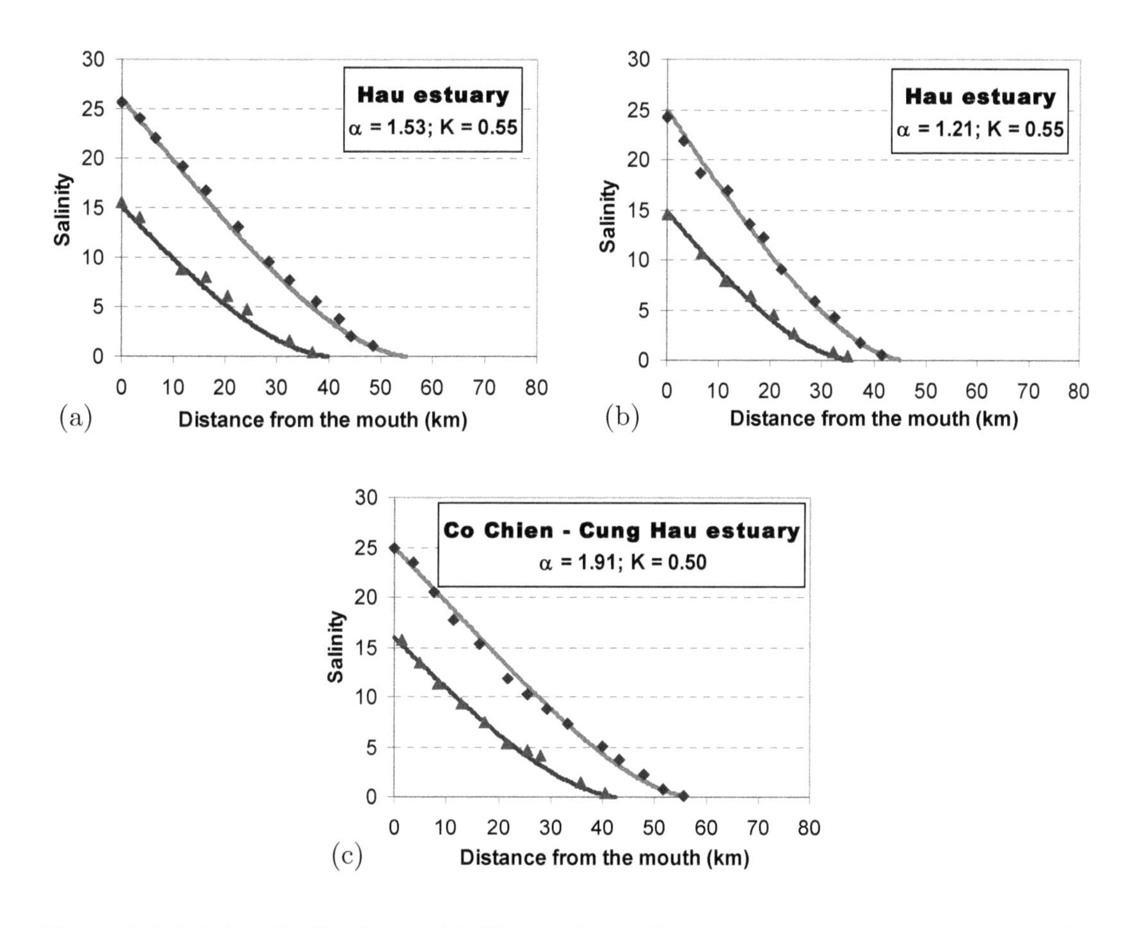

Figure 3.7 Salinity distribution of (a) The combined Hau estuary on 8 and 9 April, 2005; (b) The combined Hau estuary on 21 and 22 May, 2005; and (c) The combined Co Chien – Cung Hau estuary on 21 and 22 April, 2005, showing values of measured salinity at HWS (diamonds), and LWS (triangles), and calibrated salinity curves at HWS (upper curves) and LWS (lower curves).

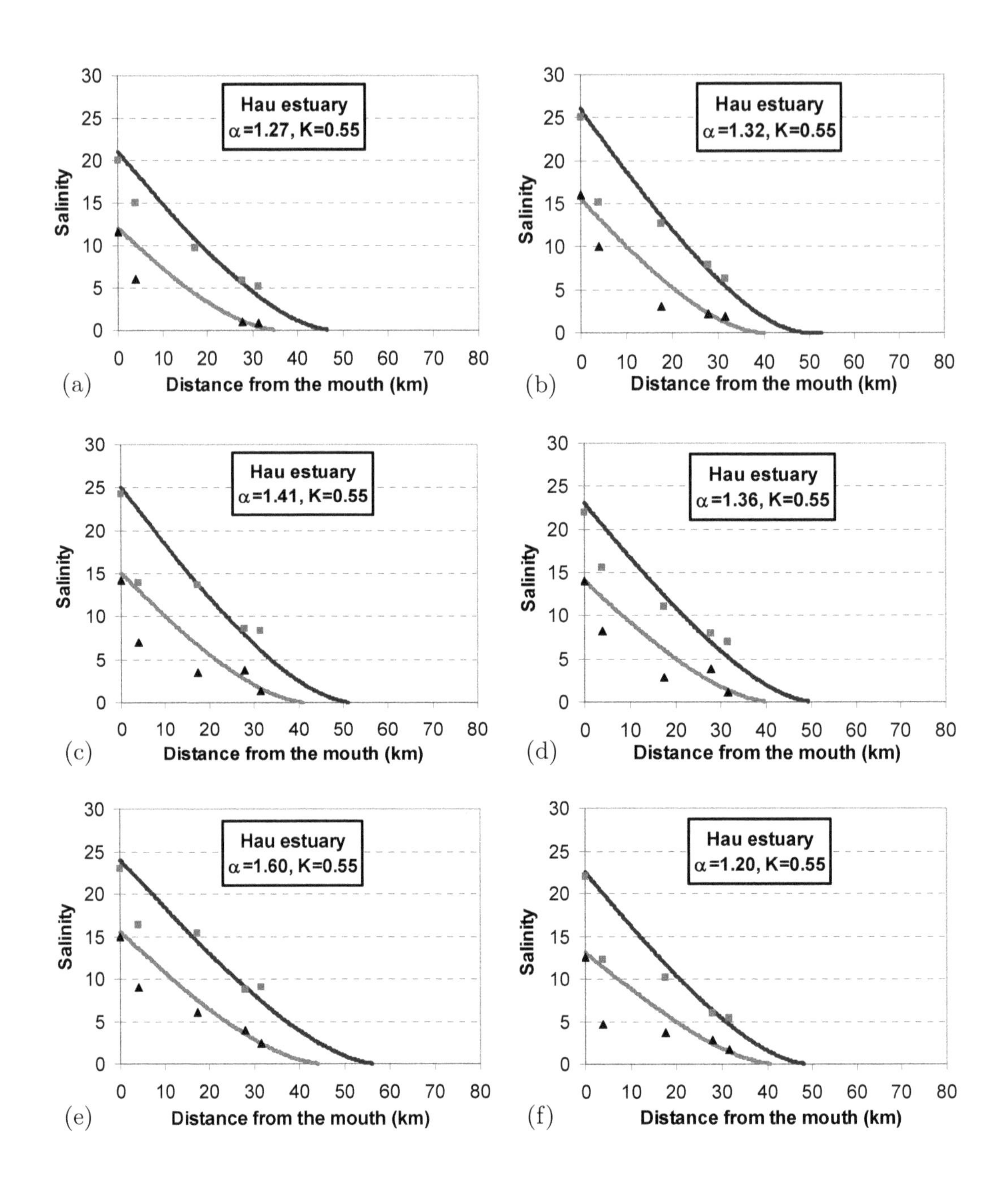

Figure 3.8 Validation results of the Hau estuary on: (a) 01 March 1998; (b) 05 April 1998; (c) 07 April 1998; (d) 02 March 1999; (e) 20 March 1999 and (f) 16 April 1999, showing values of observed salinity at HWS (diamonds) and LWS (triangles), and validated salinity curves at HWS (upper curves) and LWS (lower curves).

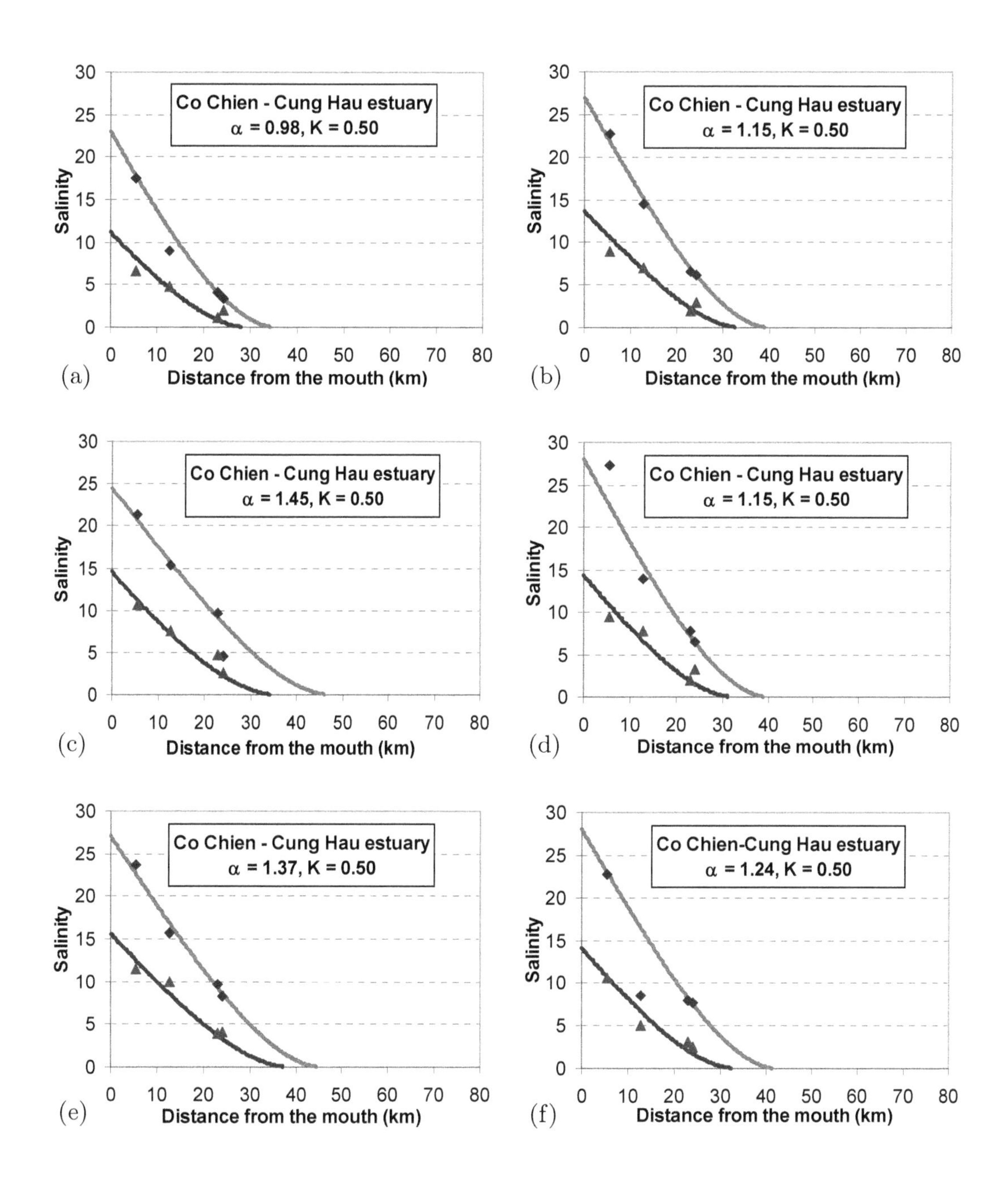

Figure 3.9 Validation results of the Co Chien – Cung Hau (combined Cung Hau and Co Chien) estuary on: (a) 16 March 1998; (b) 02 April 1998; (c) 15 April 1998; (d) 02 March 1999; (e) 21 March 1999 and (f) 19 April 1999, showing values of observed salinity at HWS (diamonds) and LWS (triangles), and validated salinity curves at HWS (upper curves) and LWS (lower curves).

3.6 PREDICTIVE MODEL

Based on the calibration and validation results of the second approach (see Section 3.5), we can use Eqs. 3.9 and 3.10 to turn the salinity model of the Mekong Delta into a predictive model. The values of K, obtained from calibration, are 0.55 and 0.50 for the Hau and the Co Chien-Cung Hau respectively, which compare fairly well with predicted values of 0.42 and 0.45, which are computed by Eq. 3.10. Similarly, the calibration values of D_0^{HWS} ($D_0^{HWS} = \alpha_0^{HWS} Q_f$, where Q_f is known from the discharge ratio provided by VNHS in combination with observations; and α_0^{HWS} is obtained through calibration against the observed longitudinal salinity distribution) for the Hau estuary and the Co Chien – Cung Hau estuary should be compared to Eq. 3.9. We notice that the calibration values of D_0^{HWS} for the measurement described in Figs. 3.8 and 3.9 fit Eq. 3.9 well (see Fig. 3.10) if we use b (i.e. the width convergence length) instead of a (i.e. the area convergence length) and if the average depth over the salinity intrusion length is taken instead of the depth at the estuary mouth.

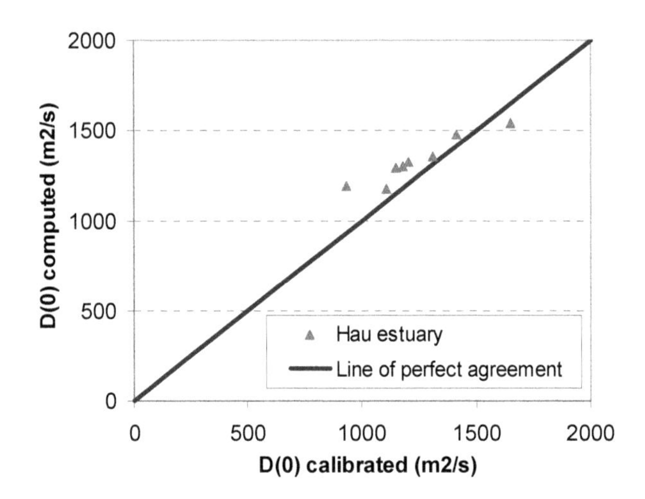

Figure 3.10 Empirical relation for D_0^{HWS} for the Hau e tuary.

For the Hau estuary and the Co Chien-Cung Hau estuary, the predictive equation then reads:

$$\frac{D_0^{HWS}}{\upsilon_0 E_0} = 1,400 \frac{\bar{h}}{b} \sqrt{N_R} \tag{3.14}$$

In Fig. 3.11, intrusion lengths computed with Eq. 3.14 are plotted together with the estuaries presented by Savenije (2005, page 171). What we observe in Fig. 3.11 is that the salt intrusion lengths at HWS of the Hau and the Co Chien-Cung Hau computed by the modified predictive model using Eq. 3.14 and Eq. 3.8 (blue triangles) plot very well within the set. To permit a good comparison, the intrusion lengths in the other estuaries have also been computed using Eq. 3.14.

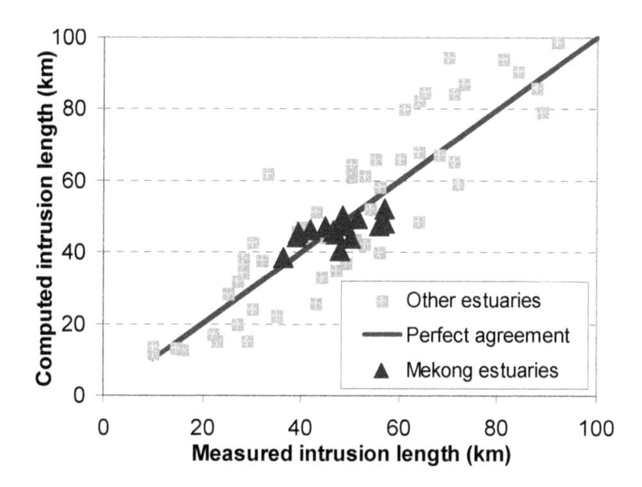

Figure 3.11 Salt intrusion length at HWS according to the modified predictive model, applied to the Mekong and other estuaries described by Savenije (2005).

3.7 DISCUSSION

One thing we have to bear in mind is that the method presented is applicable to a steady state salinity distribution. Estuaries require a certain time to adjust to changes in the boundary conditions before they reach a steady state. Savenije (2005, p. 175) found that the system response times for different estuaries are in the order of magnitude of days to months. Using Eq. 2.40, we can compute the system response time for the Mekong estuaries. The computed system response times for the Hau estuary on 8 April 2005 and 21 May 2005 and the Co Chien - Cung Hau estuary on 21 April 2005 are 7 days, 10 days and 3 days, respectively. In relation to the low variability of the discharge during the dry season, these values indicate that the estuary system is capable of adjusting itself to a new situation, and does not lag far behind an equilibrium state during the dry season. Therefore, we can use the steady state method to compute the salinity distribution in the Mekong estuaries.

The Mekong Delta has a network of fixed stations along the banks, which measure salinity values during the dry season. These stations provide hourly salinity observations. These values are not very adequate for the HWS-LWS method. For validation, we took the maximum and minimum daily values of the salinity at these stations and we assumed that these values are close to HWS and LWS values in the same day. This assumption is rather weak since the stations are generally not located near the centre of the stream but rather in the lee, often near intakes of canals.

The Van den Burgh's coefficient, K, has been taken from the calibration process from field data in 2005. These values compare fairly well with the predictive Eq. 3.10. We should realise that the predictive equation for K is still rather weak and may have to be improved in the future. Therefore, the calibrated K values for the Hau and the Co Chien – Cung Hau estuaries are considered to be the most reliable. For a detailed explanation on Van den Burgh's coefficient reference is made to Savenije (2005).

There are several limitations in this study. Firstly, there is lack of updated topographical data of the Mekong Delta. The data used in this chapter have been obtained from surveys carried out between 1991 and 1998. Later, the Mekong Committee developed a topographical database of the Mekong Delta, but this database is essentially composed of the same data with only minor modifications and updates. As we know, the Mekong Delta is morphologically active and the topography is continuously changing due to the high sediment transport capacity of the river. Hence, there still is room for improvement of the salinity model by using more recent topographical data. Secondly, the predictive model is sensitive to a number of parameters that have a certain degree of uncertainty. These are the tidal excursion E_0, the mean estuary depth $\overline{h}$ and the width convergence length b. The tidal excursion can be obtained accurately if we have good and adequate HWS and LWS salinity measurements. There is uncertainty in the determination of the average depth over the cross-section, particularly when (for instance in the Dinh An) there is a shallow part and a deep part, and when the estuary depth is not constant (there is a slight bottom slope). Similarly, there is an effect of a possible error in the convergence length due to the lack of updated topographical data. The uncertainty in the average depth may be reduced by using the analytical relations for tidal damping and wave propagation presented by Savenije (2001, 2005), Savenije and Veling (2005), but this will require additional observations of tidal damping, wave propagation and longitudinal salinity distributions. Both the uncertainties in the average depth and the convergence length are relevant for the accuracy of the predictive model.

3.8 CONCLUSIONS

In this chapter, the theory for the computation of salt intrusion in single alluvial estuaries is for the first time applied to a riverine multi-channel estuary with increasing depth in upstream direction and damped tidal excursion. Although the theory has not been developed for this situation, it is well applicable if we combine paired estuary branches and modify Eq. 3.9 into Eq. 3.14.

In view of the similar hydraulic, topographical and salinity characteristics of the branched estuaries in the Mekong, it is suggested that the multi-channel estuarine system functions as an entity and that paired branches should be considered as a single estuary branch. This procedure has been successfully applied and tested in the Dinh An and Tran De branches (the combined estuary named the Hau estuary) and the Co Chien and Cung Hau branches (the combined estuary named the Co Chien - Cung Hau estuary). Based on salinity measurements during the dry season of April and May 2005, an analytical model has been developed to compute the longitudinal salinity distribution (at HWS and LWS) for the combined estuaries, e.g. the Hau estuary and the Co Chien - Cung Hau estuary. The model has been validated with data of the dry seasons in 1998 and 1999. The overall results of salinity computation indicate that the assumption of combined branches is workable and that the simplified method can produce satisfactory results for a complex system such as the Mekong Delta.

Chapter 4

DISCHARGE DISTRIBUTION OVER ESTUARY BRANCHES

Abstract

The freshwater discharge is an important parameter for modeling salt intrusion in an estuary. In alluvial converging estuaries during periods of low flow, when salinity is highest, the river discharge is generally small compared to the tidal flow. This makes the determination of the freshwater discharge a challenging task. Even if discharge observations are available during a full tidal cycle, the freshwater discharge is seldom much larger than the measurement error in the tidal discharge. Observations further upstream, outside the tidal region, do not always reflect the actual flow in the saline area due to withdrawals or additional drainage. Discharge computation is even more difficult in a complex system such as the Mekong Delta, which is a multi-channel estuary consisting of many branches, over which the freshwater discharge distribution cannot be measured directly. This chapter presents a new analytical approach to determine the freshwater discharge distribution over the branches of the Mekong Delta by means of an analytical salt intrusion model, based on measurements made during the dry season of 2005 and 2006. It appears that the analytical approach agrees well with observations and with a hydraulic model. This chapter demonstrates that with relatively simple and appropriate salinity measurements and making use of the analytical salt intrusion model, it is possible to obtain an accurate discharge distribution over the branches of a complex estuary system. This makes the new approach in combination with the analytical salt intrusion model a powerful tool to analyze the water resources in tidal regions.

Parts of this chapter were published as:

Nguyen, A.D., Savenije, H.H.G., Pham, D.N., and Tang, D.T., 2007. Using salt intrusion measurements to determine the freshwater discharge distribution over the branches of a multi-channel estuary: the Mekong Delta case. Estuarine, Coastal and Shelf Science, doi 10.1016/j.ecss.2007.10.010.

Nguyen, A.D., and Savenije, H.H.G., 2007. New method to determine the freshwater discharge distribution over the branches of a multi-channel estuary. In D. Boyer, O. Alexandrova (Eds.), Proceedings of the Fifth International Symposium on Environmental Hydraulics, Tempe, Arizona, USA, 6pp.

4.1 Introduction

The river discharge, together with relevant parameters defining estuary shape and tidal forcing, is the key parameter determining salt intrusion in alluvial estuaries. This was demonstrated in a large number of estuaries by Savenije (1986, 1989, 1993, 2005, 2006), Brockway *et al.* (2006) and Nguyen and Savenije (2006), by the use of an analytical model. The model depends on two major drivers, the tidal variation at the estuary mouth and the freshwater discharge entering the saline area.

The determination of the freshwater discharge in estuaries is complicated, as it requires detailed measurements during a full tidal cycle. Moreover, in the dry season when the salt intrusion matters most, the magnitude of the freshwater discharge is small compared to the tidal flow (often within the measurement error of the tidal flow). It is even more difficult to determine the discharge in a complex system such as the Mekong Delta, which consists of eight branches over which the freshwater discharge is distributed (see Fig. 3.2).

The Mekong Delta has been subject to a number of studies including: Wolanski *et al.* (1996, 1998), Nguyen *et al.* (2000), Tang (2002), Wolanski and Nguyen (2005), Le (2006) and Le *et al.* (2007). These publications focus on the sediment dynamics of the delta (Wolanski *et al.*, 1996, 1998), and the flow and transport regime during the flood season (Nguyen *et al.*, 2000; Le *et al.*, 2007). In the past, several hydraulic models have been developed over time to simulate the hydrodynamic regime of the Mekong river system. Although these models were not developed for the purpose of determining the discharge distribution over the branches of the Mekong Delta, they can indeed provide this information. However, it appears that the results from these models do not always agree due to the different topographical dataset (due to the changes in the Mekong over years) and different modeling objectives (Le, 2006).

In Chapter 3, we present the development of the predictive analytical model for salt intrusion in the Mekong Delta in Vietnam, which can be used to predict the salinity distribution in the Mekong branches if topography, tide and river discharge are known. The reverse also applies: if the salinity distribution in the Mekong is known, we can estimate the river discharge.

This chapter presents a new analytical approach to determine the discharge distribution over the branches of the Mekong by means of the salt intrusion model developed in Chapter 3, based on dry season data of 2005 and 2006. These results will be compared with the results obtained by a hydraulic model, both for accuracy and efficiency.

4.2 The Mekong Delta, Vietnam

The Mekong Delta has already been introduced in Chapters 1 and Chapter 3. Here we focus on the freshwater discharge characteristics. In addition, the shape of two other main branches of the delta (i.e. My Tho and Ham Luong) is presented. The data set for this chapter is also briefly described.

4.2.1 Freshwater discharge and topographical characteristics of the Mekong Delta

In the dry season, there are two main sources for the freshwater: (i) from the main Mekong river and (ii) from the Great Lake (TonleSap) in Cambodia via the Bassac river. The total freshwater discharge in the dry season is in the order of 2,000 m^3/s, which distributes unequally over the eight branches. Observations further upstream, outside the tidal region, are available. There are several discharge measurement stations located on the main Mekong river (e.g. Kratie and Kompong Cham, about 330 km from the Dai mouth), the Bassac river (e.g. Bassac Chaktomouk, 240 km from the Dinh An mouth) and on several tributaries (e.g. Prek Kdam on the Tonle Sap river or Vam Nao on the Vam Nao river). Inside the tidal region in Vietnam, there are four discharge stations: Tan Chau and My Thuan on the Tien (Mekong) river, which are located 200 km and 100 km from the Dai mouth, respectively; and Chau Doc and Can Tho on the Hau (Bassac) river, which are located 190 km and 80 km from the Dinh An mouth, respectively (See Fig. 3.2). However, the distribution of the discharge over the branches downstream of the river system depends on a complex interaction of topography, tide, network layout (hydraulic structures, canals, etc.) and additional withdrawals or drainage. Therefore, it is difficult to obtain a reasonable estimate of the discharge distribution over the branches of the system. There are two discharge stations located in the tidal region (i.e. My Thuan and Can Tho on the Tien and Hau main branches, respectively). However, based on observations from these two stations, it is not possible to obtain a reliable estimate of the distribution of the freshwater discharge. The reason is that the freshwater discharge, which is in the order of 1,000 m^3/s, is probably smaller than the measurement error in the tidal discharge, which is in the order of 12,000 m^3/s. It would be costly to carry out detailed measurements during a full tidal cycle in the downstream end of the Mekong's eight branches and we would face a similar problem related to the measurement error.

The estuary branches Dinh An and Tran De, Co Chien and Cung Hau have the characteristics of paired estuary branches (see Chapter 3). The Tieu and Dai may also be considered as paired, although we should realize that the length of the Tieu branch is slightly larger than that of the Dai (34.5 km vs. 32.5 km). The Ham Luong branch is a single branch estuary. The Ba Lai branch, which is closed by a tidal barrier at its mouth, is not taken into account in this study because it is relatively small and it no longer is a natural branch. The estuarine characteristics of the Tran De - Dinh An, the Co Chien - Cung Hau, the My Tho (combination of the Tieu and Dai) and the Ham Luong branches correspond very well with exponential functions that follow the concept of ideal estuaries (see Chapter 3, Section 3.3). The combined My Tho and the combined Hau estuary have an inflection point located 45 km from the Dai mouth and 57 km from the Dinh An mouth, respectively. Upstream from the inflection point, the shape of the estuaries also corresponds with exponential functions, and the depth is constant. The topography of the two branches is shown in Figs. 4.1 and 4.2. Combining with information from Table 3.2, then we have a complete set of the topography of the main branches of the Mekong Delta, which is presented in Table 4.1.

Table 4.1 Estuarine characteristics of seven branches in the Mekong Delta

River	Estuary	A_0 (m^2)	B_0 (m)	$\bar{h}$ (m)	a (km)	b (km)	d (km)	a_1 (km)	b_1 (km)
Hau river	Dinh An branch	18,400	3,400	7.6	100	47	89	-	-
	Tran De branch	8,200	1,500	5.5	800	800	∞	-	-
	Combined estuary	26,600	4,900	7.5	105	51	99	140	140
Co Chien river	Co Chien branch	11,100	1,600	7.6	93	71	300	-	-
	Cung Hau branch	13,200	2,500	5.7	45	40	360	-	-
	Combined estuary	24,300	4,100	6.6	69	54	250	-	-
My Tho river	Tieu branch	7,100	1,100	6.5	180	180	∞	-	-
	Dai branch	14,500	2,300	9.3	70	38	84	-	-
	Combined estuary	21,600	3,400	7.7	71	50	170	420	420
Ham Luong river	Ham Luong branch	17,000	2,800	6.1	55	55	∞	-	-

Note: The values for the width, depth and cross-sectional area were measured at Mean Sea Level (MSL). a_1 and b_1 are the area and width convergence lengths after the inflection point, respectively.

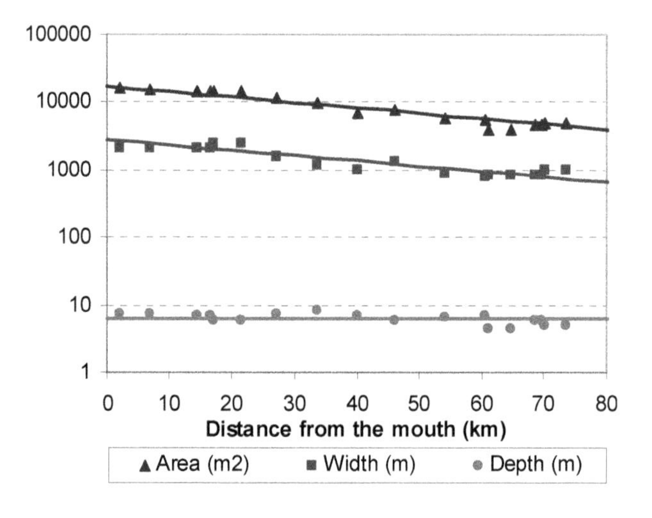

Figure 4.1 Shape of the Ham Luong estuary, showing the measured area (triangles), measured width (squares) and measured depth (circles).

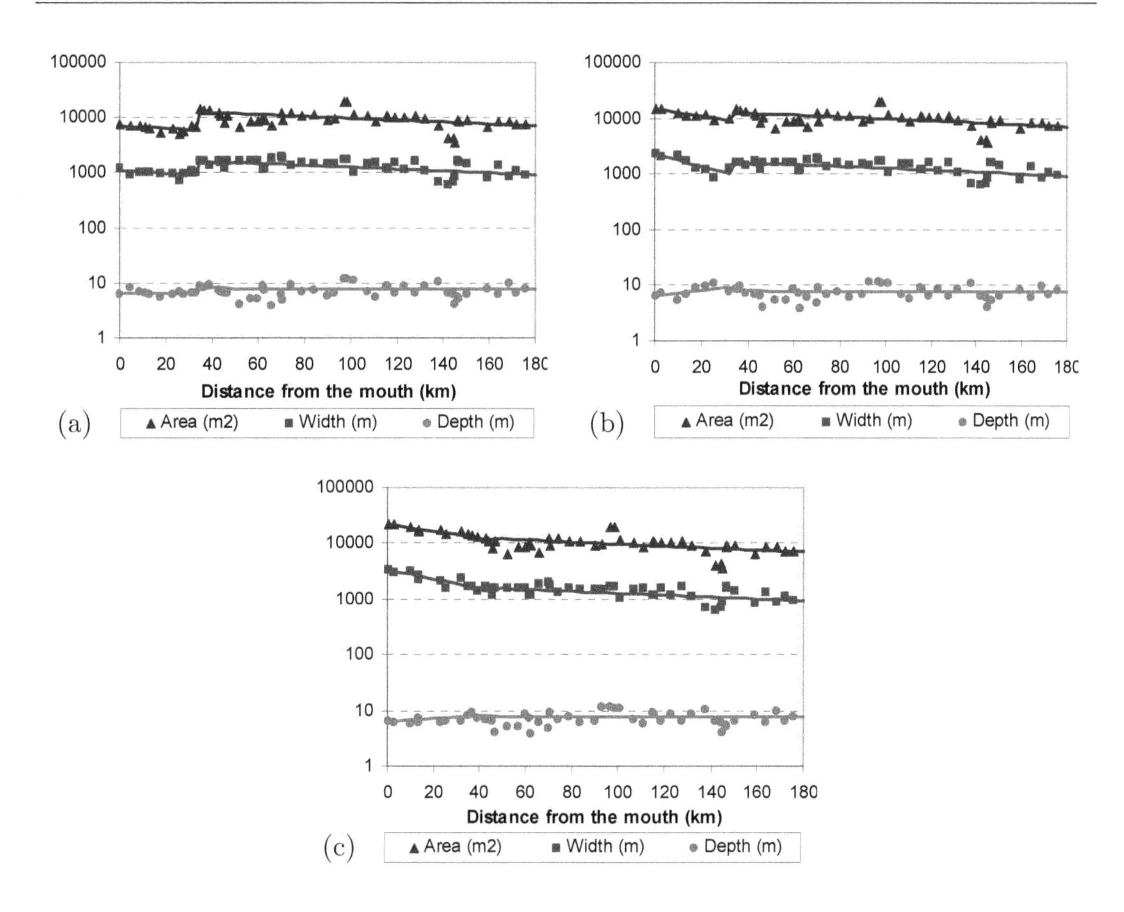

Figure 4.2 Shape of the individual branches: (a) Tieu and (b) Dai. (c) Shape of the My Tho (combination of Tieu and Dai) estuary, showing the measured area (triangles), measured width (squares) and measured depth (circles).

4.2.2 Description of the Mekong Delta's data set

On the Tien river, there are two discharge stations located at Tan Chau and My Thuan. On the Hau river, there are also two discharge stations located at Chau Doc and Can Tho (see Fig. 3.2). Normally, the discharge at Tan Chau is three to five times larger than the discharge at Chau Doc (Le, 2006). One special thing is that just 30 km downstream of Tan Chau and Chau Doc, there is a connecting river called Vam Nao, which conveys water from the Tien to the Hau river. As a result, the discharge ratio between the Hau and the Tien rivers after the Vam Nao changes substantially. There is one discharge station on the Vam Nao connecting river, however we were not able to obtain the discharge data from there. Downstream from My Thuan and Can Tho, there are no more discharge stations.

The river discharge and tidal data during the 2005's and 2006's field measurement periods were provided by the Vietnamese National Hydrometeorology Services (VNHS). Salinity data of the Mekong Delta used in this chapter consist of three sets: (i) the first set is from field measurements carried out by the author during the dry season of 2005; (ii) the second data set is from field measurements carried out by the author during the end of the dry season of 2006; and (iii) the third data set is obtained from the network of fixed stations near intakes and quays, which measure salinity values during the dry season at hourly intervals. The first set has been presented in Chapter 3, Section 3.4. The other data sets are presented here. During the moving boat measurements in 2005 and 2006, we measured the vertical salinity distribution at several points in the Dinh An, Tran De, Co Chien and Cung Hau branches. It appeared that these branches were partially-mixed and well-mixed estuaries according to the estuarine stratification classification (see Section 2.2.4).

4.3 ESTIMATION OF THE DISCHARGE DISTRIBUTION OVER THE MEKONG BRANCHES

4.3.1 Previous studies determining the discharge distribution over the Mekong branches

The recent book of Le (2006) on the salinity intrusion of the Mekong Delta provides an overview of the discharge distribution over the river system. Because it is difficult to obtain a reasonable estimate of the discharge distribution, Le (2006) discussed several hydraulic models of the Mekong Delta, which were developed over the last four decades. Although, they were not developed for the purpose of determining the discharge distribution over the branches of the Mekong Delta, they can be used to assess the discharge distribution over the river system. These models include:

(i) The NEDECO model, developed in 1974.

(ii) The Vietnamese National Hydrometeorology Services (VNHS) model, developed in 1984.

(iii) The SALO89 model, used by NEDECO in 1991.

(iv) The model developed by Nguyen Van So in the combination with the observed data in 1992.

(v) The VRSAP model, developed in 1993.

Estimates of the discharge distribution over the branches of the Mekong Delta, based on the data of these five models, are presented in Table 4.2. One can see that the discharge distribution ratio over the branches of the Mekong is not the same for the different models. The main reason is that these models were developed at different times, therefore the topographical data are different due to the changes that have taken place over years. Moreover, these models were developed to satisfy different purposes (i.e. water balance - the first and second model; irrigation - the fourth model; and salinity intrusion - the third and fifth model); therefore the choices of boundary conditions and hydraulic parameters were not the same.

It is noted that the results from the first two models (i.e. NEDECO 1974 and VNHS 1984) are almost identical. The first two models are relatively simple models and they did not sufficiently take into account the flows through the inland channel system. The results of the next three models show more or less the same pattern. It is noted that in the SALO89 and VRSAP model, the total discharge ratio of the Tien and Hau river below the Vam Nao connection is not 100%. This can be explained by the fact that before approaching the Vam Nao river, a certain amount of water flows into the inland channel system.

Table 4.2 Discharge distribution in the Mekong river system (after Le (2006, p. 43))

Model name	Discharge computed (m^3/s) (*)	Tien river below Vam Nao (%)	Hau river below Vam Nao (%)	Co Chien (%)	Cung Hau (%)	Dinh An (%)	Tran De (%)
NEDECO 1974	2,385	51.0	49.0	13.0	15.0	28.0	21.0
VNHS 1984	1,926	55.0	45.0	13.0	18.0	27.0	18.0
SALO89 1991	2,274	43.6	54.4	11.8	7.8	25.6	24.3
Nguyen Van So 1992	-	-	-	11.0	12.0	19.0	16.0
VRSAP 1993	2,280	49.7	44.3	10.9	4.5	18.2	18.0

(Continued)

Model name	Ba Lai (%)	Ham Luong (%)	Tieu (%)	Dai (%)	Others (**) (%)
NEDECO 1974	0.0	15.0	2.0	6.0	0.0
VNHS 1984	0.0	17.0	1.0	6.0	0.0
SALO89 1991	1.6	13.6	5.2	2.0	8.1
Nguyen Van So 1992	1.0	14.0	1.5	6.0	19.5
VRSAP 1993	0.1	8.7	2.3	8.4	28.9

(*) Total discharge of both the Tien and the Hau rivers, upstream of the Vam Nao connection.
(**) Internal (inland) canal system.

4.3.2 Analytical equation for estimating freshwater discharge on the basis of salinity measurements

The analytical equation is developed on the basis of the salt intrusion model presented in Chapter 3. In the reverse mode, using the same model, if the salinity distribution is known, we can use the model to predict the river discharge.

From Eq. 3.6, for the HWS situation, we have:

$$Q_f = -\frac{D_0^{HWS}}{\alpha_0^{HWS}} \tag{4.1}$$

Substituting Eq. 3.6 and Eq. 3.14 into Eq. 4.1, we obtain:

$$Q_f = -\frac{D_0^{HWS}}{\alpha_0^{HWS}} = -\frac{1400\frac{\bar{h}}{b}\sqrt{N_R}\left(\upsilon_0 E_0\right)}{\alpha_0^{HWS}}$$

with $N_R = -\dfrac{\Delta\rho}{\rho}\dfrac{ghQ_fT}{A_0E_0\upsilon^2}$ (where Q_f is negative since the positive x-axis points upstream). Hence:

$$Q_f = -\frac{1400\dfrac{\bar{h}}{b}\sqrt{-\dfrac{\Delta\rho}{\rho}\dfrac{g\bar{h}Q_fT}{A_0E_0\upsilon_0^2}}\left(\upsilon_0E_0\right)}{\alpha_0^{HWS}} = -\frac{1400\dfrac{\bar{h}}{b}\sqrt{-\dfrac{\Delta\rho}{\rho}gT\dfrac{\bar{h}}{A_0}E_0Q_f}}{\alpha_0^{HWS}}$$

or:

$$Q_f = -\left(\frac{1400\dfrac{\bar{h}^{\frac{3}{2}}}{b}\sqrt{\dfrac{\Delta\rho}{\rho}gT\dfrac{E_0}{A_0}}}{\alpha_0^{HWS}}\right)^2 \tag{4.2}$$

The notations have already been introduced in Chapter 2 and Chapter 3.

4.4 Using the analytical salt intrusion model to compute the discharge distribution

4.4.1 Salinity distribution in the Mekong branches in the dry season of 2005

There are two important parameters that have to be known in order to determine Q_f with Eq. 4.2: (i) α_0^{HWS} and (ii) E_0. The longitudinal salinity curves calibrated against measurements during HWS and LWS can provide us with accurate estimates of both parameters. α_0^{HWS} can be obtained through calibration, and E_0 is the distance between the HWS and LWS curves.

In order to obtain the theoretical longitudinal salinity curves, we have to calibrate the salt intrusion model against observations. There are two calibration parameters, i.e. K and α_0^{HWS}. A first estimate of K can be obtained by the predictive equation (i.e. Eq. 3.10 in Chapter 3). However, due to the large uncertainty of this predictive equation, the K estimate should be refined on the basis of salinity measurements. K, Van den Burgh's coefficient, is a 'shape factor' influencing the shape of the salt intrusion curve (Savenije, 1993a). K particularly determines the shape of the toe of the salt intrusion curve. Most importantly, K is not dependent on time but purely dependent on topography and tidal characteristics. α_0^{HWS} plays a small role in the shape of the salt intrusion curve but the main role in the intrusion length. The calibration process has been carried out on the basis of a quantitative procedure employing a weighted Chi-Square method with special weight to the toe of the salinity intrusion curves. It is particularly important to give emphasis to the toe of the intrusion curve because it incorporates all errors incurred in the integration of the differential equation (i.e. Eq. 3.8 in Chapter 3), which uses the downstream boundary salinity (i.e. at the mouth) as the input. The toe represents the total intrusion length, which is the key outcome of the salt intrusion model.

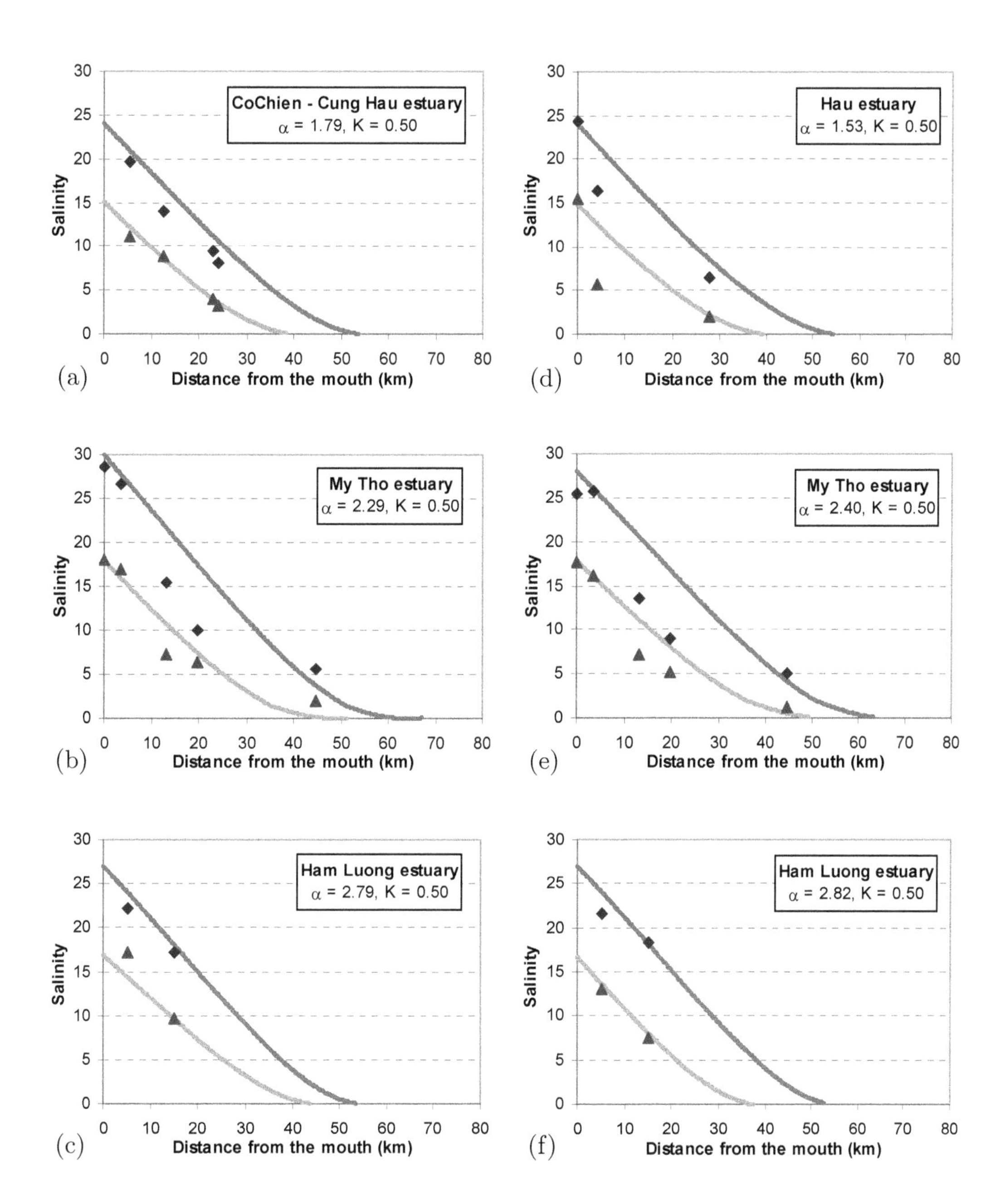

Figure 4.3 Salinity distribution of individual branches and combined branches of the Mekong: On 8 and 9 April 2005: (a) Co Chien - Cung Hau, (b) My Tho and (c) Ham Luong. And on 21 and 22 April 2005: (d) Hau, (e) My Tho and (f) Ham Luong, showing values of measured salinity at HWS (diamonds), and LWS (triangles), and calibrated salinity curves at HWS (upper curves) and LWS (lower curves).

The weighted Chi-Square method reads:

$$X_W^2 = \sum_{i=1}^{n} \left[\frac{\left[(x_i - y_i) \right]^2}{y_i} \times \left| \frac{y_0 - y_i}{y_0} \right| \right] \tag{4.3}$$

where x_i and y_i are the observed and simulated values (spatial series). y_0 is the simulated value at the mouth of the estuary (boundary condition) and n is the sample size or number of events. It is required that X_W^2 is minimized.

The quantitative procedure for salt intrusion model calibration can be expressed as the two following steps:

(i) Basing on the observed salinity curves and applying a quantitative analysis of the Chi-Square Statistic method as well as a weighted Chi-Square method with special attentions on the toe of the curves, we established a fixed value of K for each estuary branch.

(ii) Using the Chi-Square method and the weighted Chi-Square method, we quantitatively determined the α_0^{HWS} values.

The calibration results on 8 and 9 April 2005 for the Hau branch and on 21 and 22 April 2005 for the Co Chien–Cung Hau branch were presented in Chapter 3, therefore we do not repeat them here. The salt intrusion length in the Hau estuary during the dry season of 2005 was in the order of 50 km. This agrees with the value observed by Wolanski *et al.* (1998) during the dry season of 1996.

On the same dates, we used routine measurements to obtain a full set of salinity distribution in the remaining branches of the delta (see Fig. 4.3). These measurements are parts of standard measurements taken at fixed locations along the estuary, often at intakes and well beside the main current. As a result, these observations sometimes underestimate the HWS salinity, or they are affected by land drainage. This is clearly visible in the Hau, at 4 km from the mouth, where the gauge is located in an inlet. This makes the calibration results less reliable, but we feel that they are still useful for our purpose.

4.4.2 The freshwater discharge distribution in the Mekong Delta during the dry season of 2005

On the basis of the observed salinity distribution in the Mekong Delta branches, the freshwater discharge distribution has been calculated using Eq. 4.2. There are two approaches to determine the discharge distribution over the estuary branches:

(i) Approach 1: Using the parameters obtained with the salinity distribution in the individual branches, combined with the estuary shape of each individual branch, we are able to compute the discharge in each individual branch. This approach is only applicable for the Dinh An, Tran De, Cung Hau and Co Chien branch, where we have sufficient salinity measurements.

(ii) Approach 2: Using the estuary shape of the combined estuary branches, together with the parameters from the salinity distribution, we can compute the freshwater discharge of the combined branches (i.e. Hau, Co Chien - Cung Hau and My Tho).

Table 4.3 presents the computation of the freshwater discharge over the Mekong branches, using both approaches. For reasons of simplicity, here and in the following, we use the absolute value of the river discharge, which has a negative value since the positive x-axis points upstream.

Table 4.3 Freshwater discharge values in 2005 in the Mekong branches computed by means of the salt intrusion model.

Date	River	Estuary	α_0^{HWS} (m^{-1})	E_0 (km)	Q_f (m^3s^{-1})	Percentage of observations (%) (*)
8 and 9 April 2005	Hau	Dinh An branch	2.39	16.0	649	27.4
		Tran De branch	4.48	16.5	273	11.5
		Combined Hau estuary	1.53	16.5	922	38.9
	Co Chien	Combined estuary	1.79	16.0	435	18.3
	My Tho	Combined My Tho estuary	2.29	19.0	657	27.7
	Ham Luong	Ham Luong branch	2.79	18.0	219	9.2
21 and 22 April 2005	Hau	Combined Hau estuary	1.53	16.0	894	42.1
	Co Chien	Co Chien branch	3.83	16.5	189	8.9
		Cung Hau branch	4.12	16.5	183	8.6
		Combined estuary	1.91	16.5	394	17.5
	My Tho	Combined My Tho estuary	2.40	18.0	567	26.7
	Ham Luong	Ham Luong branch	2.82	17.5	208	9.8

(*) Total observed discharge of both the Tien (Tan Chau station) and the Hau river (Chau Doc station), about 30 km upstream of the Vam Nao connection.

One can see that in the case of the Tran De branch, of which the downstream shape is almost constant (i.e. a very large convergence length), we are not able to use Eq. 4.2 to compute the discharge value. This is a disadvantage of the first approach when applied to branches with near constant cross-sections. It is remarked that in Table 4.3, the discharges of the Hau, Co Chien–Cung Hau and My Tho (i.e. three combined branches) are computed by the second approach, while the discharges of the Co Chien, Cung Hau and Dinh An branch are computed by the first approach. One can see that the values computed by the second approach and the first approach for the Co Chien–Cung Hau are identical. This, once again, implies that the paired estuaries function as an entity.

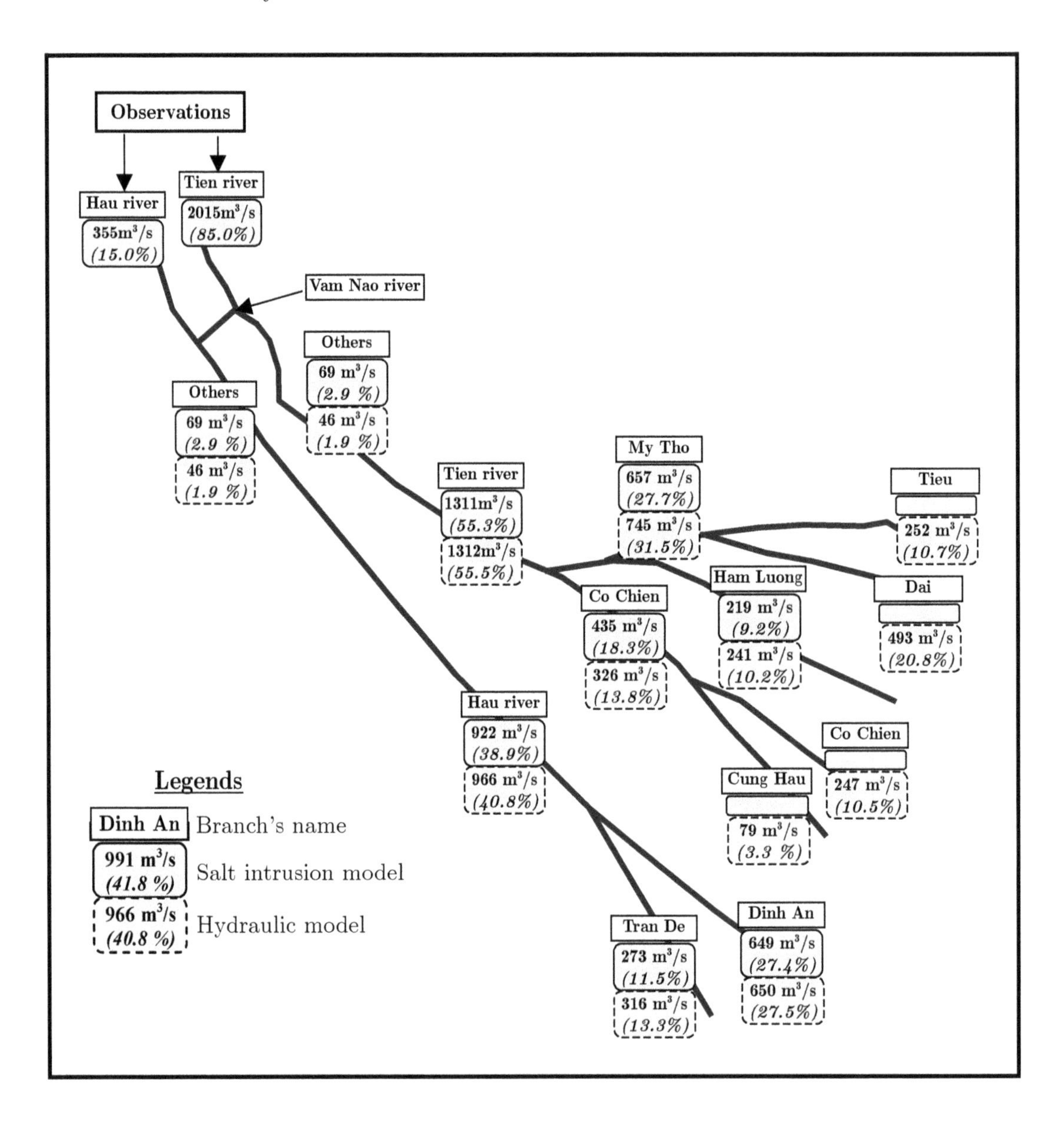

Figure 4.4 Discharge distribution over the branches of the Mekong Delta computed on the basis of the salinity observations carried out on 8 and 9 April, 2005.

The value of the Tran De branch is obtained by subtracting the values of the combined Hau estuary and the Dinh An branch. On 8 and 9 April 2005, the salinity data were not sufficient in the Cung Hau, Co Chien, Dai and Tieu branch in order to get accurate parameters for the discharge computation in each individual branch. Similarly, on 21 and 22 April 2005, data were not sufficient to obtain the right parameters for the Dinh An, Tran De, Dai and Tieu branch.

Table 4.3 shows the discharge distribution over the system and Fig. 4.4 presents an overview of the discharge distribution over the estuary branches on 8 and 9 April 2005. The following conclusions can be drawn:

(i) The discharge ratio between the Dinh An and the Tran De is in the order of 70%/30%.

(ii) The discharge ratio between the Co Chien and the Cung Hau is in the order of 50%/50%.

(iii) The discharge ratio between the combined Hau and the combined Co Chien–Cung Hau is in the order of 70%/30%.

(iv) If we assume that the value of the "others" component, that is the amount of river flow that feeds the inland canal network (see Fig. 4.4), can be split equally between the Hau and the Tien river, then we see that the discharge ratio between the Hau and the Tien river after the Vam Nao connection is in the order of 42%/58%. This agrees reasonably well with the estimation of VNHS (unpublished) on the basis of their own hydraulic model evaluation and discharge measurements in 1997 and 2000 at Tan Chau, Chau Doc and Vam Nao, which indicated that the ratio between the Hau and the Tien river after the connecting Vam Nao river is about 45%/55%.

4.4.3 The freshwater discharge distribution in the Mekong Delta at the end of the dry season of 2006

On 10 and 11 June 2006, field measurements were carried out, using the moving boat method described in Chapter 3, Section 3.4. The advantage of this measurement approach is that observations can be carried out on the same day in the paired estuary branches (i.e. Dinh An and Tran De, Co Chien and Cung Hau). Unfortunately, due to a limited number of measurement devices and boats, we could not manage to do measurements in the Tieu, Dai and Ham Luong branches as well. Therefore, routine measurements at fixed locations have been used to obtain the salinity profile of the My Tho and Ham Luong branch.

We measured the vertical salinity distribution at several points in the Dinh An, Tran De, Co Chien and Cung Hau branches. It appeared that these branches were partially mixed. This is understandable since the discharge is relatively large compared to the dry period. The total discharge of the Tien and Hau river on 10 and 11 June is in the order of 6,000 (m^3/s) compared to a discharge of 2,000 (m^3/s) in the dry season of 2005. The actual total discharge of the Tien and Hau river on 10 and 11 June 2006 provided by the VNHS was 6,276 (m^3/s).

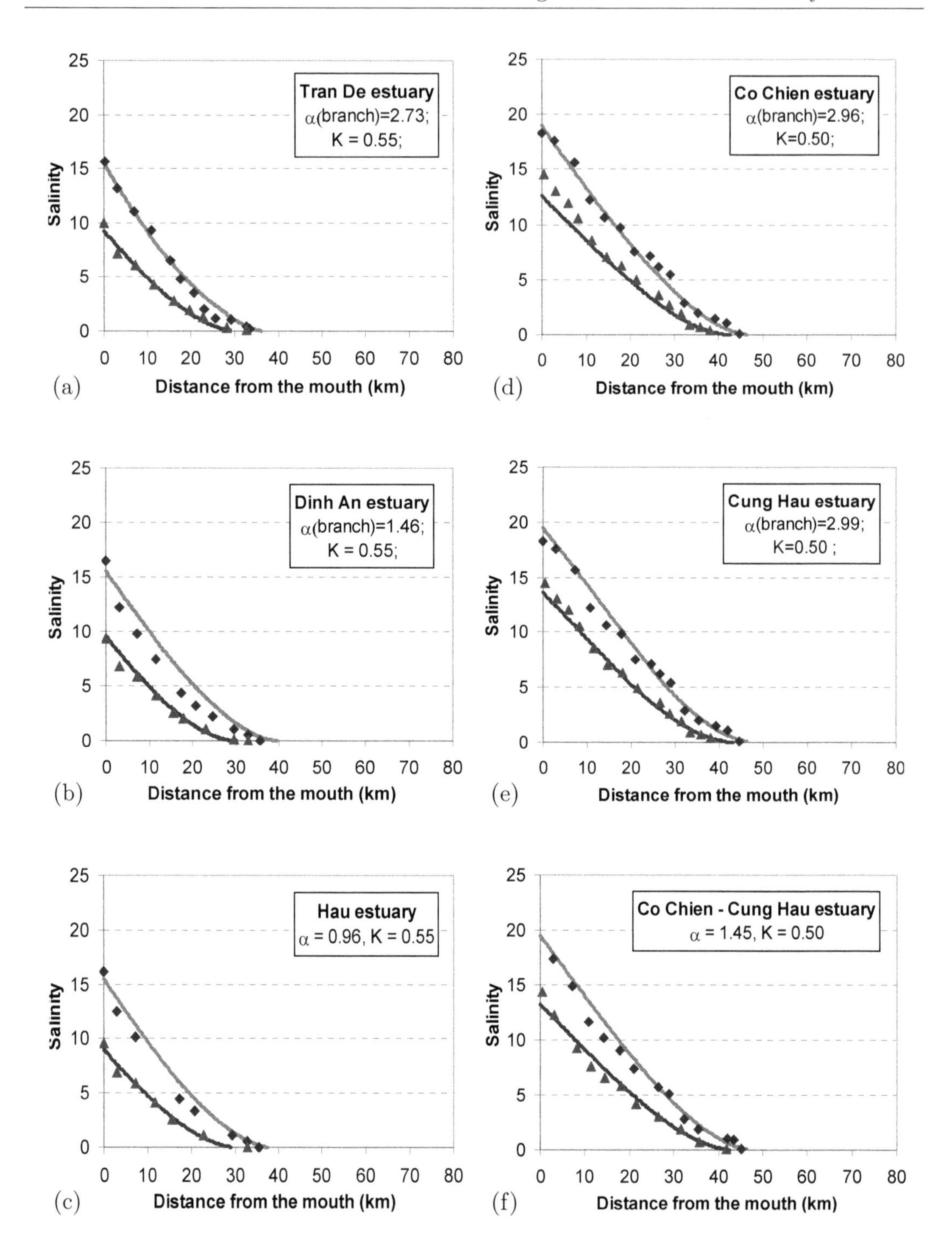

Figure 4.5 Salinity distribution of individual branches and combined branches on 11 and 12 June 2006 in the Mekong: (a) Tran De, (b) Dinh An, (c) Combined Hau estuary, (d) Co Chien, (e) Cung Hau and (f) Combined Co Chien - Cung Hau estuary, showing values of measured salinity at HWS (diamonds), and LWS (triangles), and calibrated salinity curves at HWS (upper curves) and LWS (lower curves).

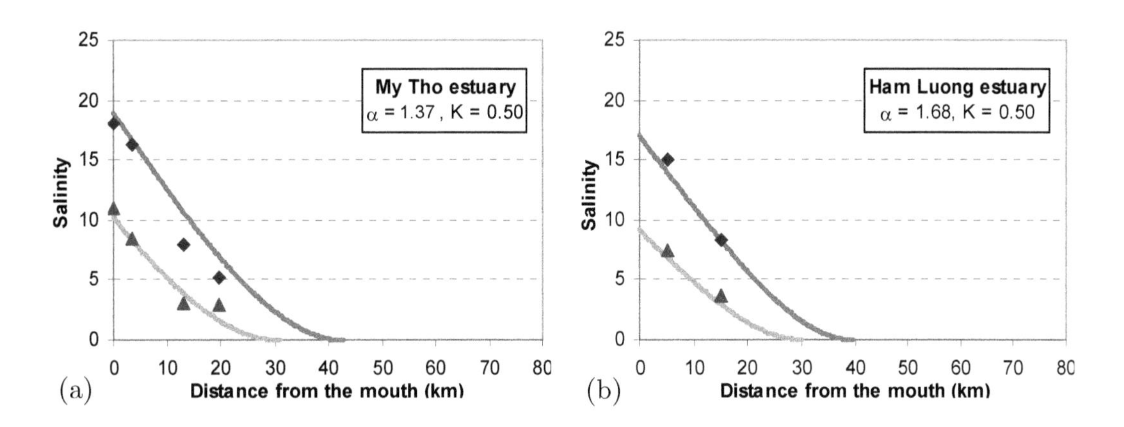

Figure 4.6 Salinity distribution of individual branches and combined branches of the Mekong on 11 and 12 June 2006: (a) My Tho and (b) Ham Luong, showing values of measured salinity at HWS (diamonds), and LWS (triangles), and calibrated salinity curves at HWS (upper curves) and LWS (lower curves).

Table 4.4 Freshwater discharge values in 2006 in the Mekong branches computed by means of the salt intrusion model.

Date	River	Estuary	α_0^{HWS} (m^{-1})	E_0 (km)	Q_f (m^3s^{-1})	Percentage of observations (%) [(*)]
10 and 11 June 2006	Hau	Dinh An branch	1.46	10.5	2,283	36.4
		Tran De branch	2.73	11.0	840	13.4
		Combined Hau estuary	0.96	11.0	3,123	49.8
	Co Chien	Co Chien branch	2.96	11.0	423	6.7
		Cung Hau branch	2.99	11.0	463	7.4
		Combined estuary	1.45	11.0	911	14.5
	My Tho	Combined My Tho estuary	1.37	14.0	1,353	21.6
	Ham Luong	Ham Luong branch	1.68	13.5	453	7.2

[(*)] Total observed discharge of both the Tien (Tan Chau station) and the Hau river (Chau Doc station), about 30 km upstream of the Vam Nao connection.

The calibration results, based on measurement data of 10 and 11 June 2006, are presented in Figs. 4.5 and 4.6. Table 4.4 shows the computation of the freshwater discharge over the Mekong branches on 10 and 11 June 2006.

From the computed results on 8-9 April 2005 and 10-11 June 2006 in the Hau river; and 21-22 April 2005 and 10-11 June 2006 in the Tien river, one can conclude that the discharge values computed by the two approaches (i.e. the single branch and the combined branch approach) agree fairly well for the computed days. This, again, implies that the paired estuaries function as an entity.

One can see in Table 4.4 that: (i) the discharge ratio between Dinh An and Tran De is 73% / 27%; (ii) the discharge ratio between Co Chien and Cung Hau is 48% / 52%. This agrees well with what we have seen in Section 4.4.2. However, we can also see that the discharge ratio between the Hau and Co Chien–Cung Hau is 78%/22% and the discharge ratio between the Hau and Tien river is 53%/47%. Apparently the discharge ratio in the Hau is larger at the start of the wet season. This may also be because of errors in the salinity measurements, which were carried out under difficult (rainy and windy) weather conditions.

Hence, we can conclude that the salinity intrusion regime and the discharge distribution pattern in the Mekong can be well described by the analytical salt intrusion approach over a wide range of river discharge.

4.5 USING A HYDRAULIC MODEL TO COMPUTE THE DISCHARGE DISTRIBUTION

4.5.1 Schematization

The new approach to use the salinity measurements to estimate the discharge distribution is tested against the results of a hydraulic model that uses the observed upstream discharge as the upstream input and the observed tidal variation at the downstream boundary. For this purpose, we have used the MIKE11 model. For detailed information on the MIKE11 package, readers are referred to: http://www.dhigroup.com/Software/WaterResources/MIKE11.aspx.

The data for the schematization of the hydraulic model, including topographical and infrastructure data, have been obtained from the 1998 and 2000 version of the Mekong river system schematization developed by the Mekong River Commission in the ISIS software. Some parts of the estuarine topography of the Dinh An, Tran De, Cung Hau, Co Chien, Dai and Tieu branches have been obtained from the data of the Southern Institute for Water Resources Research in Vietnam. Figure 4.7 presents the schematization of the Mekong Delta in MIKE11. The model has been calibrated and validated based on data of the dry seasons in 2005 and 2000. As an indication of model performance, we only show the calibration results of April 2005 at the Can Tho station (80 km from the Dinh An mouth) in Fig. 4.8.

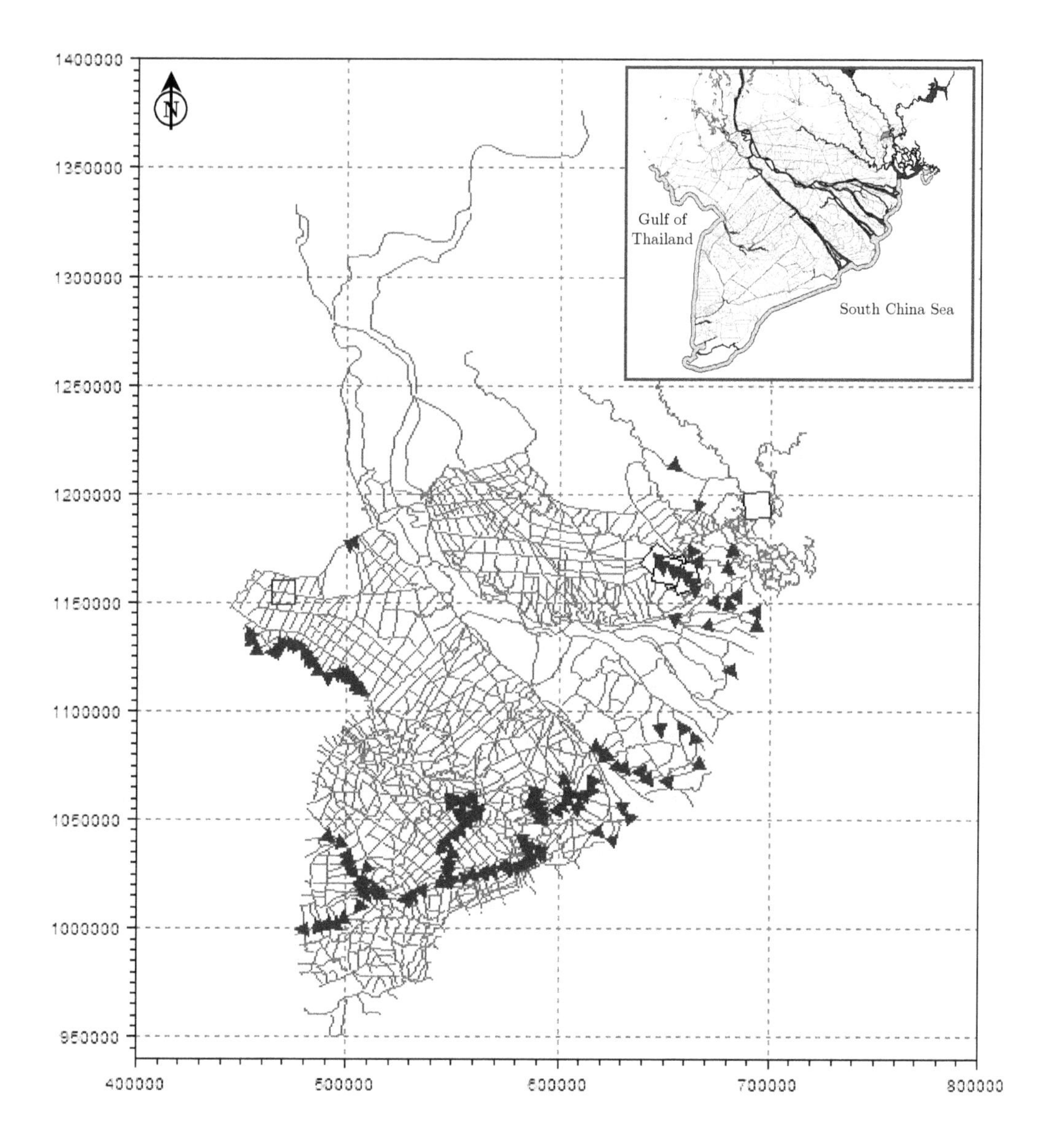

Figure 4.7 The Mekong Delta schematization in MIKE11, showing the available intake structures and sluice gates in triangles.

4.5.2 Freshwater discharge distribution

Based on the model results, we are able to derive the freshwater discharge values in the branches of the Mekong Delta. The computed results are presented in Table 4.5, also as a percentage of the observed upstream river discharge. Figure 4.9 shows these results compared to the discharge distribution computed with the salt intrusion model. In addition, Fig. 4.4 presents an overview of the discharge distribution over the branches on 8 and 9 April 2005.

It is noted that the sum of the discharge values of all branches is not necessary equal to the total discharge. This is understandable since a certain amount of water can enter or leave the inland channel system and a certain amount of water can go into the Ba Lai branch.

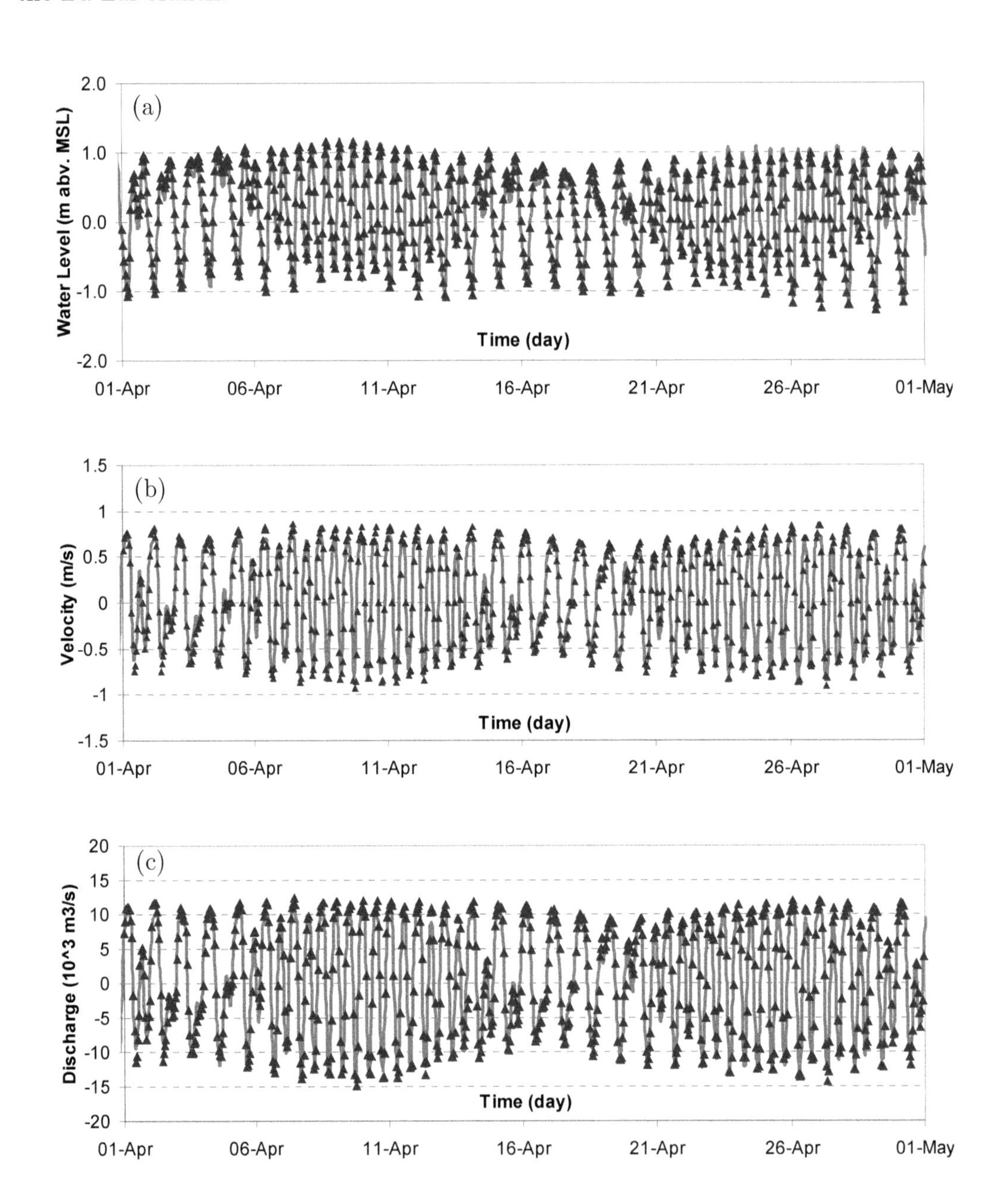

Figure 4.8 Hydrodynamic model results for the Can Tho station (80 km from the Dinh An mouth): (a) water level; (b) velocity; and (c) discharge, showing modelled values (drawn lines) and observed values (triangles).

Table 4.5 Discharge distribution above and below the Vam Nao river (computed by the hydraulic model)

Date	Observations(*) (m^3/s) & (%)	Total discharge upstream VN (m^3/s) & (%)	Tien river upstr.VN (m^3/s) & (%)	Hau river upstr. VN (m^3/s) & (%)	Tien river downstr. VN (m^3/s) & (%)	Hau river downstr. VN (m^3/s) & (%)
8 and 9 April 2005	2,370 (100)	2,197 (92.7)	1,684 (71.1)	513 (21.6)	1,312 (55.5)	966 (40.8)
21 and 22 April 2005	2,122 (100)	2,191 (103.3)	1,682 (79.3)	509 (24.0)	1,100 (52.1)	934 (44.0)

(Continued)

Date	Co Chien branch (m^3/s) & (%)	Cung Hau branch (m^3/s) & (%)	Dinh An branch (m^3/s) & (%)	Tran De branch (m^3/s) & (%)	Tieu branch (m^3/s) & (%)	Dai branch (m^3/s) & (%)	Ham Luong (m^3/s) & (%)	Total (m^3/s) & (%)
8 and 9 April 2005	247 (10.4)	79 (3.3)	650 (27.5)	316 (13.3)	252 (10.7)	493 (20.8)	241 (10.2)	2,278 (96.3)
21 and 22 April 2005	218 (10.3)	95 (4.5)	616 (29.1)	318 (15.0)	212 (10.0)	409 (19.4)	166 (7.9)	2,034 (96.1)

(*) Total observed discharge of both the Tien (Tan Chau station) and the Hau river (Chau Doc station), about 30 km upstream of the Vam Nao (VN) connection.

4.6 DISCUSSION

4.6.1 Comparison between the salt intrusion and the hydraulic model

Figure 4.9 compares the discharge distribution obtained by the salt intrusion model and the hydraulic model for the two different survey dates (see also Tables 4.3 and 4.5). It appears that: (i) the results of the two models agree reasonably well, especially in the determination of the discharge of the paired branches; (ii) the discharge ratio between the Tien and Hau river after the Vam Nao connection is in the order of 58%/42%; and (iii) there are no substantial changes in the discharge distribution over the branches between 8-9 April and 21-22 April 2005, although some differences occur in the Ham Luong, My Tho and the inland canal system.

The good agreement between the hydraulic model and the analytical model confirms the validity of the new analytical approach for determining the discharge distribution of the Mekong. In addition, the joint consideration of the two models reduces the uncertainty of the predictions. The comparison with the hydraulic model is no proof of the correctness of the new method, but it certainly is a strong indication of the applicability of the analytical model.

The MIKE11's application is in fact a solution method for the 1-D mass and momentum balance equation. The flow distribution determined by MIKE11 depends on the boundary conditions and the topography. These have been taken from the official database of the Mekong river commission (the project on Water Utilization Programme - WUP), which is the most reliable source so far. A 2-D hydraulic model could be useful and could possibly provide a better solution because the topography at the separation points would be more realistic. However, considering the data set of the Mekong Delta, it is not yet possible to develop such a model for the entire delta.

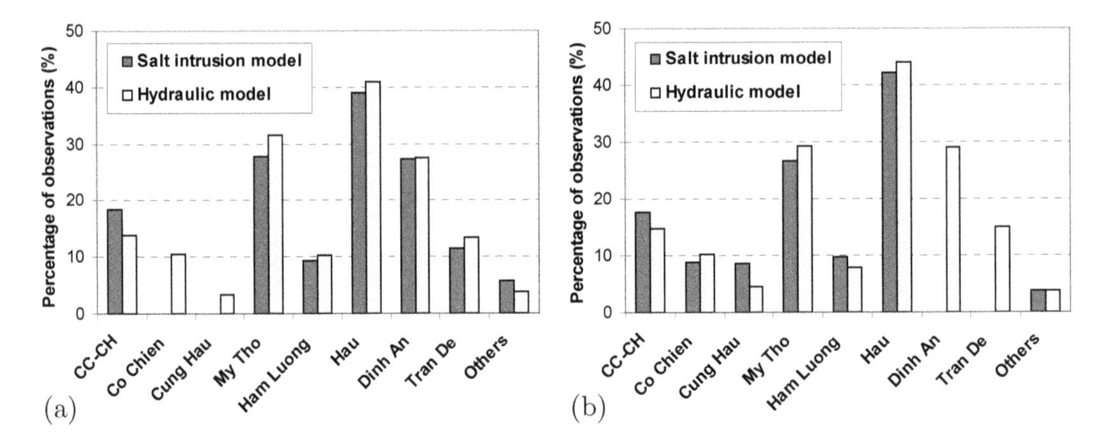

Figure 4.9 Comparison between the salt intrusion model and MIKE11: (a) on 8 and 9 April, 2005 and (b) on 21 and 22 April 2005. (Note: CC-CH: Co Chien - Cung Hau)

4.6.2 Comparison of the salt intrusion model with other models

Table 4.6 and Fig. 4.10 compare the discharge distribution over the branches of the Mekong obtained by the salt intrusion model with other hydrodynamic models. One can notice that the total observed discharges of all models are in the order of 2,000 (m^3/s) computed over one month (from April to May), which is in the middle of the dry season. For reasons of comparison, the salt intrusion model's data and MIKE11's data have also been averaged using the data of 8, 9, 21 and 22 April 2005.

It is remarked that the results from the first two models (i.e. NEDECO 1974 and VNHS 1984) are almost identical. The first two models are relatively simple models and they did not sufficiently take into account the possible inland water flows.

The results of the first three models show more or less the same pattern and they differ from the salt intrusion model results. The sixth model (i.e. MIKE11) is the most updated hydraulic model, giving a good agreement with the salt intrusion model, as we have discussed earlier. However, comparing to the five remaining models, it seems that the sixth model and the salt intrusion model overestimate the discharge of the My Tho branch. The third, fourth and fifth model provide a mixed pattern with high ratios on the inland water flows.

In Table 4.6 and Fig. 4.10, we see significant changes in the three systems of paired estuaries (i.e. My Tho, Co Chien - Cung Hau and Hau). If we may assume that the models adequately describe the situations of the Mekong Delta at the time when they were developed, then we can see an interesting shift in the discharge distribution over four decades in the Mekong Delta, whereby parts of the water flow in the Cung Hau - Co Chien, the Ham Luong and the Hau are conveyed to the My Tho (see Fig. 4.10 for the ovals and arrows). If this is indeed true, then the question arises: what the reason of the shift may have been. There are two possible explanations: (i) morphological changes; and (ii) the development of the inland channel system and channel structures.

Table 4.6 Freshwater discharge ratio (in %) in the Mekong branches: Comparison between the salt intrusion model and other models.

No	Name of model	CC–CH (%) (*)	*Co Chien* (%)(**)	*Cung Hau* (%)(**)	My Tho (%)	Ham Luong (%)	Hau (%)	*Dinh An* (%)(**)	*Tran De* (%)(**)	Others (%)(***)
1	NEDECO 1974	28.0	13.0	15.0	8.0	15.0	49.0	28.0	21.0	0.0
2	VNHS 1984	31.0	13.0	18.0	7.0	17.0	45.0	27.0	18.0	0.0
3	SALO89 1991	19.6	11.8	7.8	7.2	13.6	49.9	25.6	24.3	9.7
4	Nguyen Van So 1992	23.0	11.0	12.0	7.5	14.0	35.0	19.0	16.0	20.5
5	VRSAP 1993	15.4	10.9	4.5	10.7	8.7	36.2	18.2	18.0	29.0
6	MIKE11 2005	14.3	10.4	3.9	30.4	9.0	42.4	28.3	14.2	3.8
7	Salt intrusion model	17.9	9.1	8.8	27.2	9.5	40.5	28.5	12.0	4.8

(*) CC-CH: The Co Chien – Cung Hau combined branch.
(**) Italic names are individual branches.
(***) Others (including the Ba Lai branch and the inland channel system).

As we know, the Mekong Delta is morphologically active and the topography is continuously changing due to the high sediment transport capacity of the river. It is expected that there have been some changes over the branches of the Mekong for the last four decades and they may have led to the adjustment of the discharge distribution over the branches. Some changes in the water discharge distribution may have occurred due to the development of the inland channel system. Considering the quick development of the inland channel system for irrigation and navigation, especially during the period from 1990 to 2000, it is expected that a substantial amount of water flows in and out of the main branches, which may be conveyed from one branch to another. The fourth and the fifth models show an increasing trend that stopped recently. Since 1990, a large number of intake structures and sluice gates have been constructed to control the flow of the inland channel system, especially in the downstream part of the branches in order to prevent salt-water intrusion (Nguyen and Nguyen, 1999). The schematization in MIKE11 has been set-up with the recent data on the available structures and their operation. The most updated models (i.e. salt intrusion and MIKE11) show a reduction of the inland water flow compared to the fourth and fifth models. This reduction may be because the structures are closed during the dry season to prevent salt-water intrusion, and therefore they keep a certain amount of freshwater out of the inland channel system.

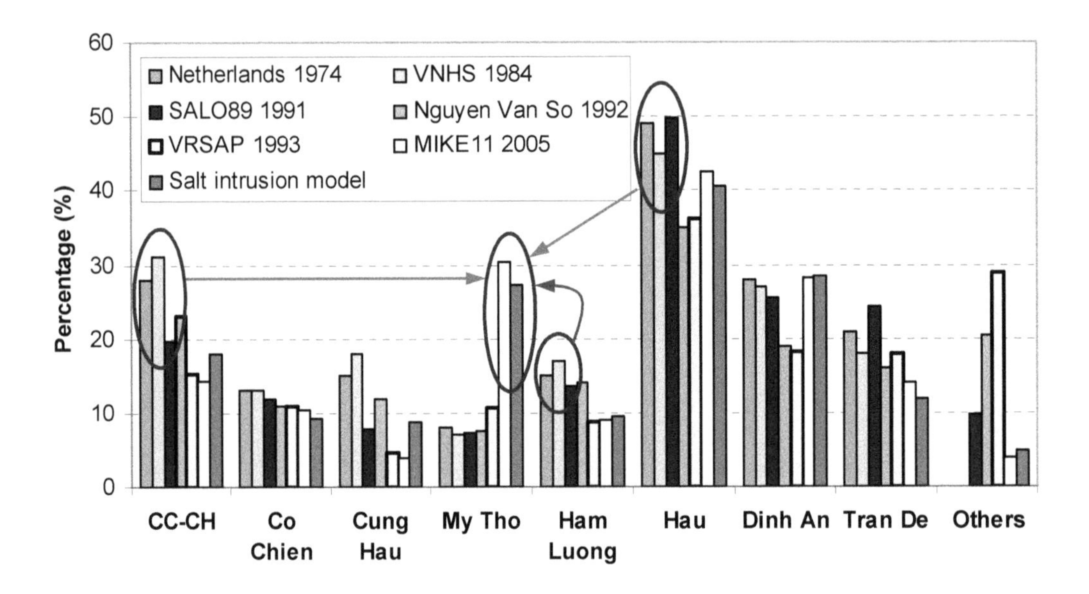

Figure 4.10 Comparison between the salt intrusion model and other models

4.6.3 Limitation of the salt intrusion model

We have seen that the salt intrusion model performs almost identically to the MIKE11 hydraulic model, which is a complex model requiring a heavy set of data and has been built on the basis of detailed topographical information and the latest knowledge about the infrastructure. This implies that a simple model can provide a good insight into a complicated system like the Mekong Delta. However, the salt intrusion model, when used to compute the discharge distribution, also has its limitations.

Firstly, it is required to carry out salinity measurements over every branch within the same period (preferably within the same day). Although these measurements provide a good overview of how salinity and discharge distribute over the branches of the Mekong, the organization of these measurements requires substantial equipment, transportation and human resources. The authors could not manage to carry out all these measurements at the same time. Therefore, the solutions given in Section 4.4.2 and 4.4.3 are not complete for the Tieu, Dai and Ham Luong branch; and instead we used salinity observations at fixed stations. The calibration results in these branches are hence less reliable.

Secondly, the predictive discharge values obtained by Eq. 4.2 are sensitive to errors in α_0^{HWS}, $\overline{h}$ and b. α_0^{HWS} is obtained from the salt intrusion model. Due to K being constant in the model, only α_0^{HWS} is sensitive to the prediction of the freshwater discharge. We can see from Eq. 4.2 that a relative error of 5% in α_0^{HWS} results in a 10% error in the discharge prediction. This may seem a large error, but a direct observation of the freshwater discharge is subject to much larger errors (see Section 4.2.1). The reasonably accurate values of $\overline{h}$ and b can be obtained due to the sufficient topographical data of the Mekong Delta branches, although we have to bear in mind that there is a lack of the most updated topographical data. In addition, as we have already discussed in Section 3.7 (Chapter 3), the uncertainty in the average depth may even be reduced by using analytical relations for tidal damping and tidal wave propagation presented by Savenije (2005) or Savenije and Veling (2005), but this will require additional observations of tidal damping and tidal wave propagation.

Thirdly, Eq. 4.2 is limited to partially to well-mixed estuaries where stratification effects are not significant and hence, where riverine flows are relatively small comparing to tidal flows. This, however, is the case in most alluvial coastal plain estuaries during the dry season (Savenije, 2005).

Finally, the salt intrusion model and the discharge distribution over the branches of the Mekong Delta are obtained under the assumption that the discharge values and ratios are constant during the considered period. In view of the fast reaction time for a change in the freshwater discharge of the Mekong Delta during the dry season, in the order of one week (see Section 3.7, Chapter 3), it is a justified assumption. However, this assumption might not be correct for estuaries with a long reaction time.

4.7 Conclusions

This chapter presents a new analytical approach to determine the discharge distribution over the branches of the Mekong Delta by means of a predictive analytical salt intrusion model. It appears that the analytical model agrees well with observations and with the results of a more complex hydraulic model. This chapter shows that with relatively simple and appropriate salinity measurements and making use of the analytical salt intrusion model, it is possible to obtain a good picture of the discharge distribution over a multi-channel estuary. This makes the new analytical approach in combination with the analytical salt intrusion model a powerful tool to analyze the water resources in tidal regions.

Chapter 5

TIDAL CHARACTERISTICS IN MULTI-CHANNEL ESTUARIES

Abstract

This chapter aims at: (i) exploring the characteristics of the tidal wave in the Scheldt estuary and the Mekong Delta; and (ii) testing the agreement between a set of analytical equations describing tidal wave characteristics, hydrodynamic models, and observations. Observations in alluvial estuaries indicate that the tidal damping in an estuary appears in one of three types: amplified, un-damped (ideal) or damped; and the phase lag between HW and HWS (as well as between LW and LWS) lies between zero and $\pi/2$. Moreover, a damped tidal wave moves slower than indicated by the classical equation for wave celerity, whereas the celerity is higher if the tidal wave is amplified. The Mekong Delta, Vietnam, is a multi-channel and riverine estuary consisting of eight branches. The tidal wave in the Mekong Delta is damped and therefore it is expected that the tidal wave moves slower than the classical wave celerity, and it has a large phase lag. Whereas the Scheldt estuary is a multi ebb-flood channel estuary located in the Netherlands and Belgium. The tidal wave in the Scheldt estuary is strongly amplified and therefore the tidal wave moves considerably faster than computed by the classical equation while it has a small phase lag. We find a good agreement between observations, the analytical equations and the hydrodynamic model computation. Moreover, based on the comparison between these three sources, comments are made in order to improve the performance of the set of analytical equations and the hydrodynamic models.

Parts of this chapter were published as:

Nguyen, A.D., Savenije, H.H.G., Pham, D.N., and Tang, D.T., 2007. Tidal wave propagation in the branches of a multi-channel estuary: the Mekong Delta case. In: V. Penchev, H.J. Verhagen (Eds.), Proceedings of the 4th International Conference PDCE 2007, Varna, Bulgaria, pp. 239-248.

Nguyen, A.D., Savenije, H.H.G., and Wang, Z.B., 2007. Tidal wave characteristics in a braided ebb-flood channel estuary. In: B. Cetin, S. Ulutürk (Eds.), Proceedings of the 8th International Conference MEDCOAST 2007, Alexandria, Egypt, pp. 1283-1294.

5.1 INTRODUCTION

A number of studies have been performed to describe the tidal dynamics in estuaries, ranging from mentioning and analyzing a number of tidal dynamic characteristics and behaviors (e.g. Pillsbury, 1956; Ippen, 1966; Parker, 1991; Cartwright, 1999; or Savenije, 2005) to investigating particular issues, such as tidal computations (e.g. Dronkers, 1964; or Parker *et al.*, 1999); interaction between river discharge and tides (e.g. Vongvisessomjai, 1987; or Horrevoets *et al.*, 2004); tidal wave propagation (e.g. Dronkers, 1964; Prandle and Rahman, 1980; Jay, 1991; Friedrichs and Aubrey, 1994; Lanzoni and Seminara, 1998; Godin, 1999; or Savenije and Veling, 2005) or tidal damping (Savenije, 1998, 2001), etc. In these publications, it appears that the characteristics of the tidal wave propagating into an estuary can be described through three main factors: tidal wave celerity, phase lag and tidal range variability (i.e. tidal damping/amplification).

The classical equation for tidal wave celerity (see Pillsbury, 1956; or Harleman, 1966) is widely used to describe the propagation of a tidal wave in estuaries. The conditions for the equation's derivation (i.e. constant cross section and no friction) do not apply in alluvial estuaries, where the cross section varies exponentially along the estuary axis and friction cannot be neglected. The tidal damping in an estuary appears in one of three types: amplified, un-damped (ideal) or damped. The phase lag between HW and HWS (as well as between LW and LWS) lies between zero and $\pi/2$. Determination of tidal damping and phase lag has received not much attention nor their effect on tidal wave celerity. Savenije *et al.* (2007) used a simple harmonic solution with the complete non-linearized Saint-Venant equations and developed a set of new equations describing these three main characters of the tidal wave in an estuary. This set of equations showed good agreement with observations.

The Mekong Delta, Vietnam (See Figs. 1.1 and 3.2) is a multi-channel and riverine estuary consisting of eight branches, which can be grouped in three paired estuary branches (see Chapter 4, Section 4.2). The tidal wave in the Mekong Delta is damped and therefore it is expected that the tidal wave moves considerably slower than computed by the classical equation. Whereas in the Scheldt estuary located in the Netherlands and Belgium (see Fig. 1.2), an amplified tidal wave occurs, which moves faster than indicated by the classical tidal wave equation.

In this chapter, firstly, the characteristics of the tidal wave propagation in the multi ebb-flood channel estuary (Scheldt) and paired estuary branches (Mekong) are analyzed. Secondly, on the basis of a set of analytical equations developed by Savenije *et al.* (2007), hydrodynamic models and observations, comparisons are made in April 2005 for the Mekong Delta and in June 1995 and June 1998 for the Scheldt estuary. Finally, on the basis of these comparisons, it appears that the set of analytical equations and the hydrodynamic models have strong and weak points. Recommendations are made to improve the performance of the hydrodynamic models and the set of analytical equations.

5.2 TOOLS FOR INVESTIGATING TIDAL WAVE CHARACTERISTICS IN ESTUARIES

5.2.1 Set of analytical equations

We have already introduced the set of analytical equations in Chapter 2, Section 2.5.2. In this chapter, three equations are used: the phase lag equation (i.e. Eq. 2.51), the damping equation (i.e. Eq. 2.53) and the celerity equation (i.e. Eq. 2.54). We shall elaborate further on the celerity equation in this section in order to cater for different moments of consideration.

For different moments during one tidal cycle (i.e. High water - HW; High water slack - HWS; Low water - LW and Low water slack - LWS), Eq. 2.54 can be further elaborated as follows:

$$\left(c_{HW} - \upsilon \sin\varepsilon\right)^2 = \frac{1}{r_s} g\left(h+\eta\right)\left(\frac{1}{1-D}\right) \tag{5.1}$$

$$c_{HWS}^2 = \frac{1}{r_s} g\left(h+\eta\cos\varepsilon\right)\left(\frac{1}{1-D}\right) \tag{5.2}$$

$$\left(c_{LW} + \upsilon \sin\varepsilon\right)^2 = \frac{1}{r_s} g\left(h-\eta\right)\left(\frac{1}{1-D}\right) \tag{5.3}$$

$$c_{LWS}^2 = \frac{1}{r_s} g\left(h-\eta\cos\varepsilon\right)\left(\frac{1}{1-D}\right) \tag{5.4}$$

The notations have already been introduced in Section 2.5.2.

The topographical data used for the analytical equations have been obtained from Haas (2007) and Savenije *et al.* (2007) for the Scheldt estuary, and from Table 4.1 in Chapter 4 for the Mekong Delta branches.

5.2.2 Hydrodynamic models

We use 1-D hydrodynamic models to analyze tidal wave characteristics in the Mekong Delta and the Scheldt estuary. The first one is the MIKE-11 model applied to the Mekong Delta and the second one is the SOBEK-RE model applied to the Scheldt estuary.

The data for the schematization of the MIKE11 model of the Mekong Delta have been obtained from the 1998 and 2000 version of the Mekong river system schematization developed by the Mekong River Commission in the ISIS software. The model has been calibrated and validated by the author on the basis of the dry season data of 2000 and 2005. For detailed information on the model and its performance, reference is made to Chapter 4, Section 4.5.

For detailed information on SOBEK-RE, readers are advised to consult the technical manual of SOBEK-Flow (Delft Hydraulics, 2006). The Scheldt estuary schematization in SOBEK-RE contains 246 network points (nodes), 270 branches and 270 cross section profiles (See Fig. 5.1). The model has been calibrated and validated based on hydrodynamic data of 1995 and 1998, respectively.

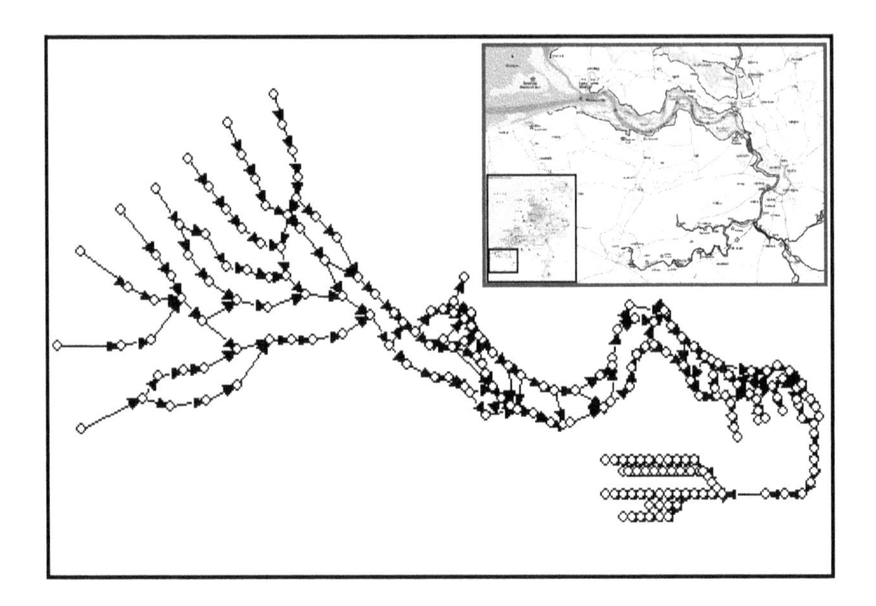

Figure 5.1 The Scheldt estuary schematization in SOBEK-RE (not to scale), showing the network nodes in circles.

5.2.3 Observations

Observations of the Mekong Delta used in this chapter consist of two sets: (i) the first set is from field measurements carried out by the authors during the dry season of 2005 and (ii) the second set is from data of a network of fixed stations along the estuaries. We carried out field measurements on the Hau river in the dry season of 2005, using the moving boat method. The second data set is obtained from the network of fixed stations near intakes and quays, which measure the water level during the dry season at hourly intervals. Unfortunately, no observations are available for the phase lag in the Mekong Delta.

Observations of the Scheldt estuary have been obtained mainly from the network of fixed water-level stations. The Scheldt estuary has a better data set compared to the Mekong Delta due to the dense station network (23 stations for the entire system with 15 stations in the Netherlands) with 10-minute intervals. No observations are available for the phase lag in the Scheldt estuary.

5.3 TIDAL WAVE CHARACTERISTICS IN MULTI-CHANNEL ESTUARIES

In this section, by a combination and comparison of three sources: observations, analytical equations and hydrodynamic models, we present the characteristics of tidal damping, phase lag and tidal wave propagation, in the two main study areas: the Mekong Delta and the Scheldt estuary. Firstly, in Section 5.3.1, the tidal wave characteristics in the Mekong Delta, which is a multi-channel and riverine estuary consisting of eight branches including three paired estuaries, are presented. Secondly, the tidal wave characteristics in the Scheldt estuary, which is a multi ebb-flood channel estuary, are analyzed in Section 5.3.2. Finally, in Section 5.3.3, these characteristics are compared to the analytical solutions of Savenije *et al.* (2007).

5.3.1 Mekong Delta

The hydrodynamic model of the Mekong river system and its results

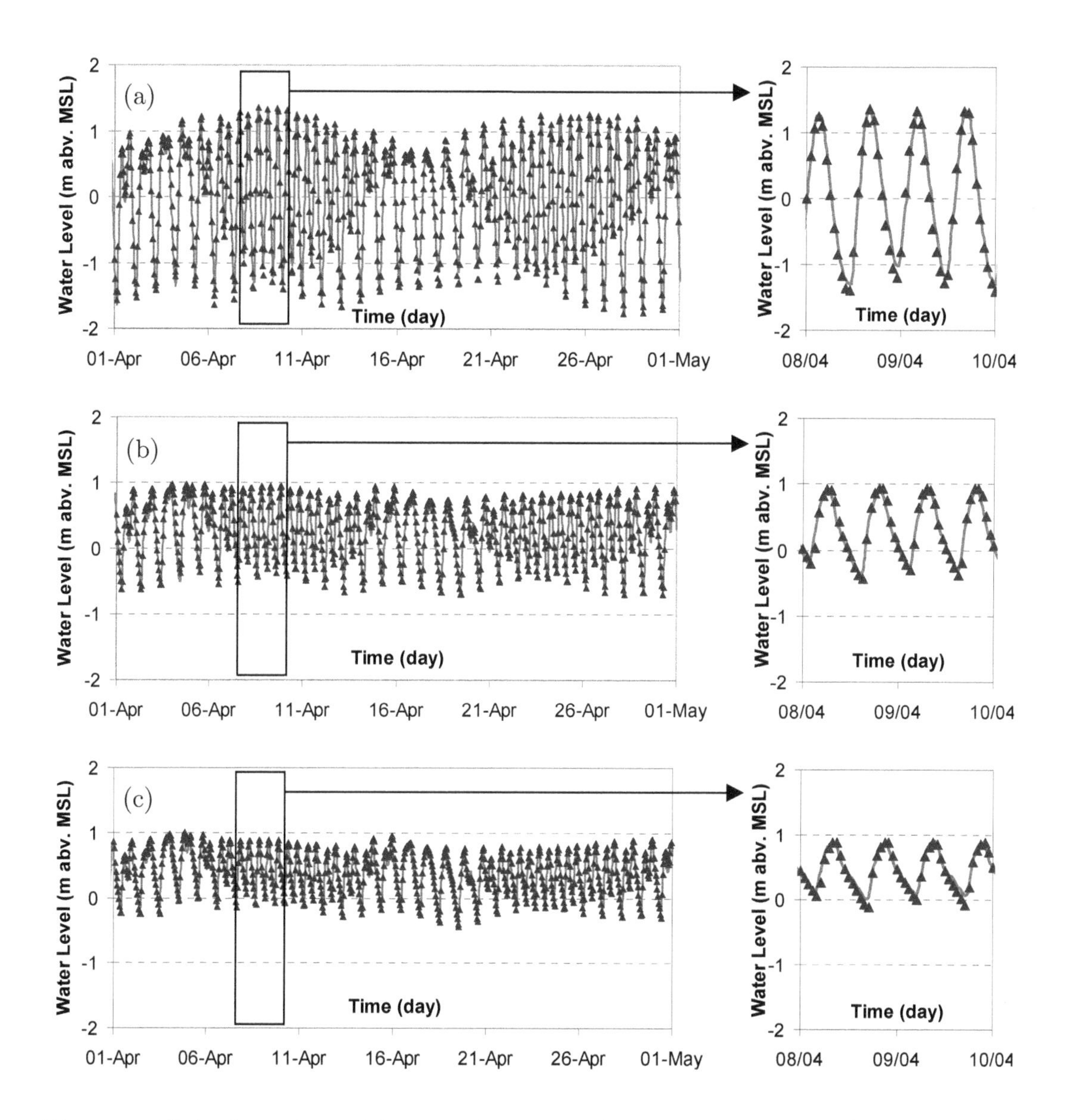

Figure 5.2 Hydrodynamic model results (drawn lines) against observations (triangles) at: (a) Dai Ngai station; (b) Long Xuyen station; and (c) Chau Doc station in the Hau river.

Along the two main branches of the Mekong system (i.e. Tien and Hau), there are a number of observation stations. They can be listed as: Vam Kenh, My Thuan, Cao Lanh and Tan Chau (0, 100, 140 and 200 km from the Dai mouth, respectively) and My Thanh, Dai Ngai, Can Tho, Long Xuyen and Chau Doc (0, 35, 80, 135 and 190 km from the Dinh An mouth, respectively) as well as several stations in the Co Chien (e.g. Tra Vinh, 25 km from the Co Chien mouth) and in the Ham Luong branch.

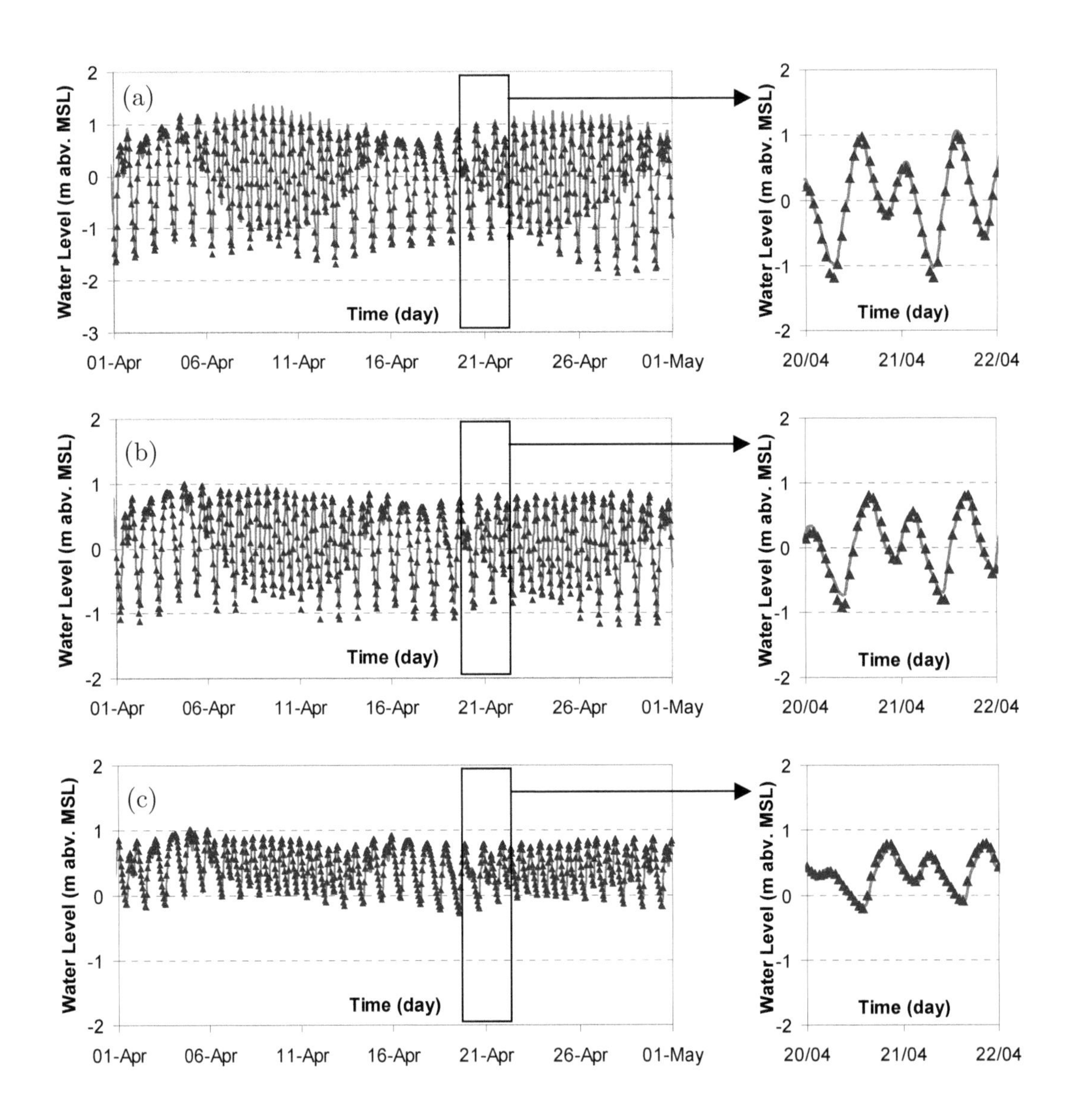

Figure 5.3 Hydrodynamic model results (drawn lines) against observations (triangles) at: (a) Tra Vinh station (Co Chien river); (b) My Thuan station; and (c) Tan Chau station in the Tien river.

As an indication of the model performance, in Chapter 4, Section 4.5, the calibration results in the Can Tho station have been presented, showing water level, discharge and velocity. Therefore, in Figs. 5.2 and 5.3, we present the observed and modelled water levels at Dai Ngai, Long Xuyen and Chau Doc on the Hau river; and My Thuan, Cao Lanh and Tan Chau on the Tien river to obtain a complete picture of the delta. Because of the one-hour interval resolution of observations, we can hardly see the difference in timing between observations and the model's results. It appears that the calibration results in the Hau river are slightly better than in the Tien river, especially in determining peaks and troughs. The peaks are generally well represented, but we see that the hydrodynamic model almost systematically underestimates the troughs of the tidal wave.

We see that the tide in the Mekong Delta branches is of the flood dominance type (Dyer, 1997). The flood phase has a shorter duration than the ebb phase. This is due to frictional distortion resulting in overtides (Speer and Aubrey, 1985; Aubrey and Speer, 1985; and Friedrichs and Aubrey, 1994).

Tidal damping and phase lag characteristics

Hau river system

Figure 5.4 presents the longitudinal development of the tidal range and the phase lag, as computed by the hydrodynamic model, for the situations of 9 and 21 April 2005. Eye-catching in Fig. 5.4 is that the tidal range and phase lag change at the location of the junction where the Dinh An and Tran De branches meet (about 35 – 40 km from the mouth). This causes a jump in the phase lag. Further upstream there are no significant jumps, although a small jump can be recognized near 80 km where the river separates into two short branches. The results of the analytical damping equation agree well with the hydrodynamic model and with observations with exception of the junctions at 35-40 km and 80 km.

Until 125 km from the mouth, the phase lag between HW and HWS is generally larger than the phase lag between LW and LWS due to the effect of overtides. Further upstream, near 140 km (on 9 April) and near 125 km (on 22 April), we can see the switch between the HW and LW phase lag. This is because of the fresh river discharge. Close to the junction point, the time lags between HW-HWS and LW-LWS are largest (i.e. about 140 minutes for HW phase lag and 120 minutes for LW phase lag). Phase lag values on 21 April (between spring tide and neap tide) are generally larger than the values on 9 April (spring tide). This can be explained by looking at Eq. 2.51, where ε is proportional to $1/c$. During spring tide on 9 April, the wave celerity is larger than on 21 April (due to the larger h). As a result, ε is smaller in spring tide. However, apparently the analytical phase lag equation does not agree well with the hydrodynamic model in the determination of the phase lag upstream of the 80 km mark. Upstream of the 80 km mark is the zone where freshwater discharge can play a significant role on the phase lag. The analytical equation does not account for this effect. Also there may be the effect of bottom slope, which has not been fully accounted for in the derivation of the phase lag equation.

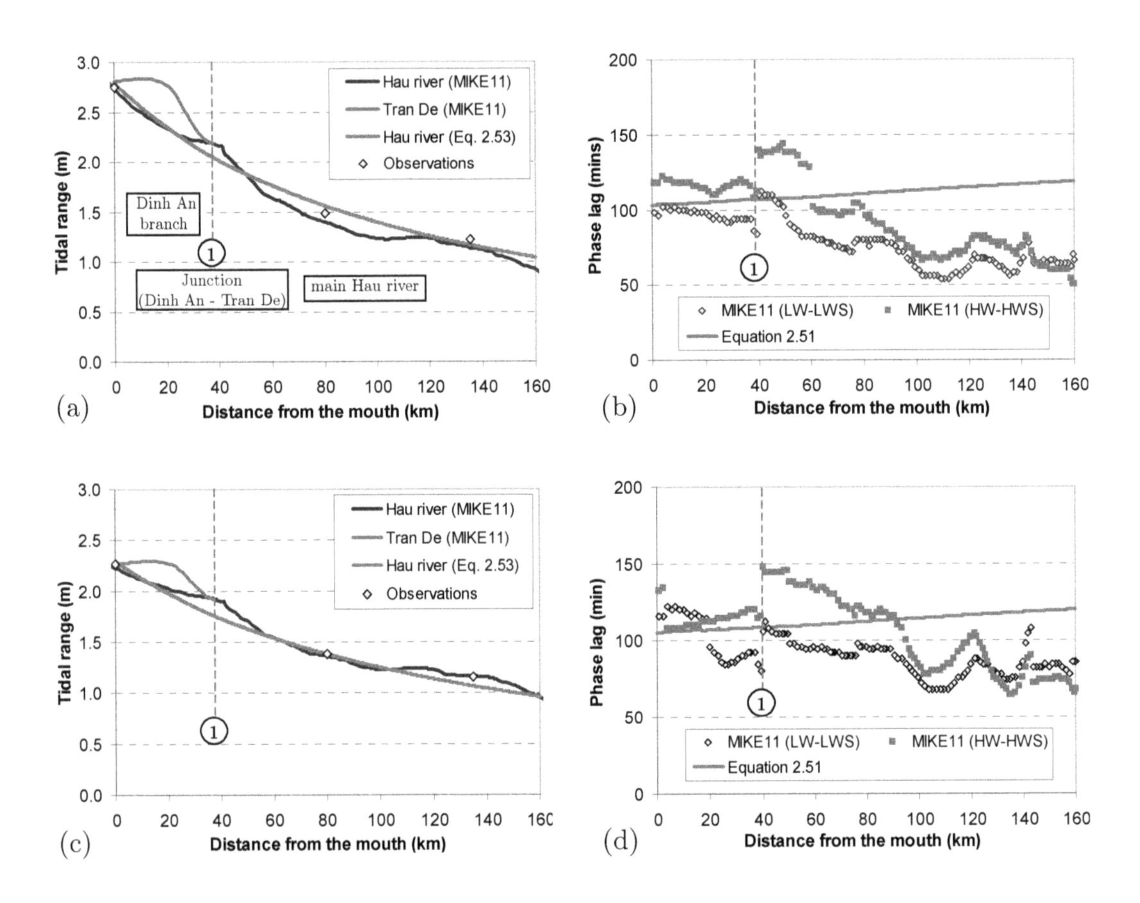

Figure 5.4 (a) Tidal damping and (b) phase lag in the Hau river on 9 April 2005 (spring tide); (c) Tidal damping and (d) phase lag in the Hau river on 21 April 2005 (weak spring tide).

Tien river system

Figure 5.5 shows the longitudinal development of the tidal range and the phase lag, as computed by the hydrodynamic model, for the situations of 9 and 21 April 2005. Also here there are some sudden and relatively large jumps in the tidal range of the Ham Luong and Co Chien river. These jumps happen at the junction point between Co Chien and Cung Hau in the Co Chien river (27 km from the mouth), and at the nose of the island, which separates the Ham Luong into two smaller branches (30 km from the mouth). Upstream and downstream of the junctions of the Ham Luong and Co Chien, we can see that the tidal range is almost constant. In the My Tho river, at the junction between the Tieu and Dai branches, there is no noticeable change in the tidal range. However, in the reach from 25 to 30 km from the Tieu mouth, there is a change in tidal range, due to the connection between the Tieu branch and one big inland channel network. The damping equation gives a very good agreement with the general trend hydrodynamic model as well as with observations, confirming the damping pattern of the wave.

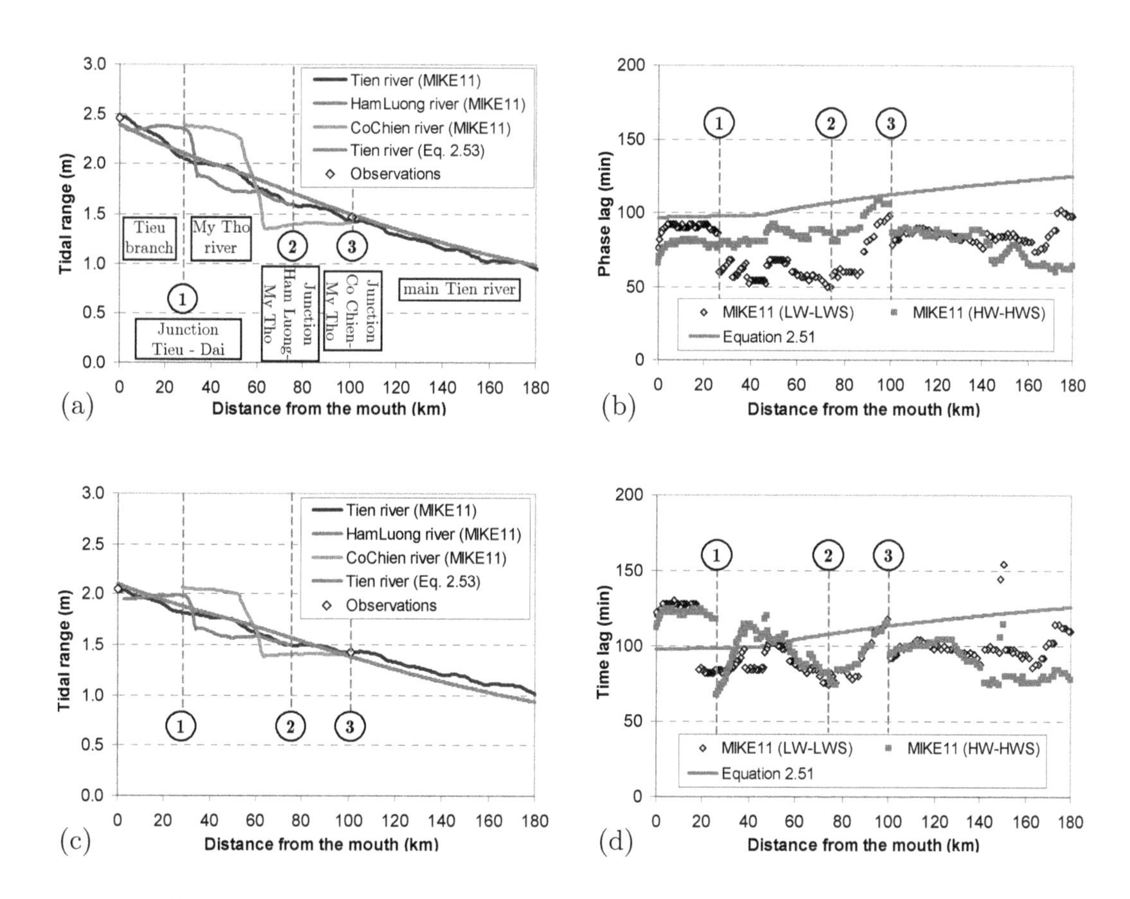

Figure 5.5 (a) Tidal damping and (b) phase lag in the Tien river on 9 April 2005; (c) Tidal damping and (d) phase lag in the Tien river on 21 April 2005.

The phase lag pattern of the Tien river system is more complicated than that of the Hau river due to the complexity of the system. Upstream of the 140 km point, the LW phase lag is larger than the HW phase lag due to the effect of the fresh river discharge. As expected, generally, phase lag values on 21 April (weak spring tide) are somewhat larger than the values on 9 April (spring tide). Similarly to the Hau, the phase lag equation does not give a good agreement with the hydrodynamic model upstream of the 100 km mark, in particular for the case on 9 April. Also here it is suspected that the slight bottom slope and the river discharge are the reason for this discrepancy.

Tidal wave propagation in the Mekong river system

Hau river system

Figure 5.6 presents a comparison on tidal wave celerity between observations, results computed from the analytical equations (Eqs. 5.2 and 5.4) and the hydrodynamic model in the situation on 9 and 21 April 2005. Because of different water levels (i.e.

different water depths), the HWS and the LWS do not have the same propagation speed (Friedrichs and Aubrey, 1988; Wang *et al.*, 2002). One can see that the classical wave celerity is much faster than the observations. This is because the classical equation does not take into account the effect of convergence and friction. The latter is dominant in damped estuaries such as the Mekong Delta. The wave celerity equation takes into account the effect of tidal damping/amplification; therefore, it is expected to better predict the propagation of the tidal wave. This expectation is confirmed in Fig. 5.6.

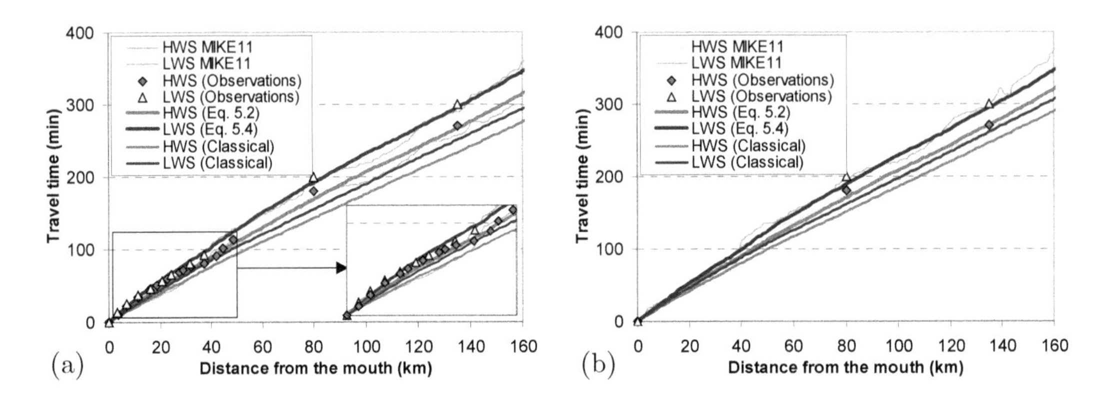

Figure 5.6 Wave propagation in the Hau river on: (a) 9 April 2005; and (b) 22 April 2005, comparison between hydrodynamic model, classical equation, analytical equation and observations.

The Hau river contains two paired branches (i.e. Tran De and Dinh An). We can see in Fig. 5.6 that there is a remarkable change in the tidal wave celerity computed with the hydrodynamic model at 35-40 km from the mouth, which is the schematised junction of the Dinh An and Tran De. The observations also reveal this change, however much smaller. The results of the analytical model do not show a significant change. We expect that there is a change in tidal damping and wave propagation patterns at the junction; however it could be smaller than what we obtain from the hydrodynamic model. The big change in the hydrodynamic model may be because of: (i) the schematised cross-sections; and (ii) lack of appropriate number for cross-sections in order to correctly simulate the transition between the paired branches and the main river.

Observations agree well with the analytical equation, taking $r_s = 1.1$ and $v = 1.3$ m/s. According to the cross-section profile, $r_s = 1.1$ is acceptable for the downstream reach and upstream reach of the estuary as we consider the dry season flow. As an artefact of the topography and schematisation imposed in the hydrodynamic model, the computed results show fluctuations at some points. However, generally, the results from the model agree reasonably well with the wave celerity equation and observations.

Tien river system

Comparing to the Hau river, the Tien river system is relatively complex, as it contains three big branches (i.e. My Tho, Co Chien and Ham Luong) and five sub-branches (i.e. Tieu, Dai, Ham Luong, Co Chien and Cung Hau). The tidal wave characteristics are complicated, as we can see in Figs. 5.5 and 5.7. The wave propagation changes dramatically at the junction points (i.e. 30 km, 80km and 100 km from the Tieu mouth). Due to the shorter length of the Co Chien, the tidal wave from the Co Chien mouth propagates landward and arrives at the junction point between the Tien and the Co Chien river earlier than the wave from the My Tho and Ham Luong river. After arriving at the junction point, it appears that water from the Co Chien moves backward into the main Tien river until it meets the peak of the wave, after which it continues upstream.

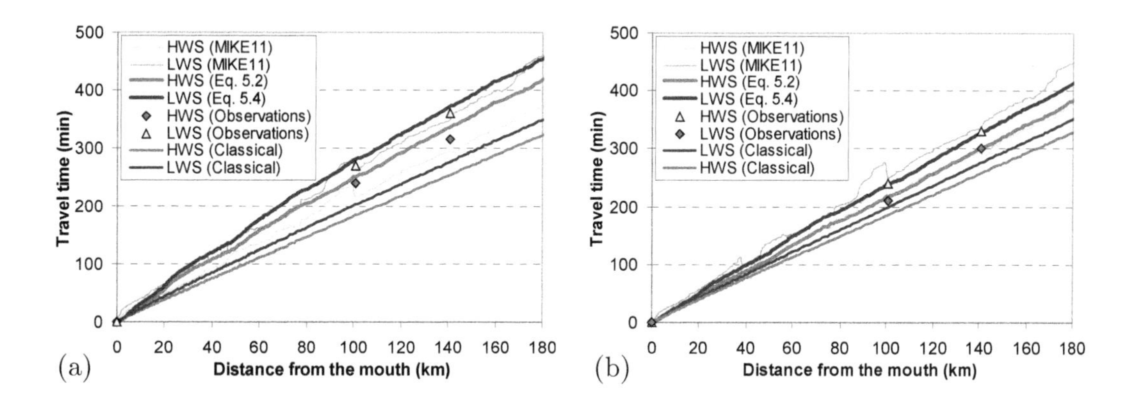

Figure 5.7 Comparison between hydrodynamic model, classical equation, analytical equation and observations in the main Tien river: (a) on 9 April 2005 and (b) on 22 April 2005.

It appears that the main Tien river (i.e. the so-called My Tho river downstream of the 100 km mark) is the dominant branch of the system. Looking at the areas, widths and depths of the branches (see Chapter 4, Section 4.2), we do not see significant differences. The Ham Luong is the smallest branch and the Co Chien is the biggest one. Looking at the length of the three main branches: Ham Luong (74 km), Co Chien (82 km) and main Tien (100 km from the Tieu mouth to the junction of Tien and Co Chien), we can see that the main Tien is the longest branch. The tidal range and wave propagation mostly follow the longest route along the Tien, as we can see in Figs. 5.5a, 5.5c and 5.7.

Generally, observations do not agree very well with the wave celerity equation nor with the model's results. On 9 April, the observations are closer to the results of the hydrodynamic model, while the analytical equation gives different results, especially at HWS. On 21 April, we obtain a better agreement between these three sources, particularly at LWS. As an artefact of the topography and schematisation in the model, the computed results show fluctuations at some points, which are similar to the case of the Hau river.

Discussion

Looking at Figs. 5.5 and 5.7, it is concluded that the longest branch (i.e. the main Tien river) determines the patterns of wave propagation and tidal range in the Tien river. Sudden jumps in tidal wave propagation occur at the junction points, due to the interaction between the branches, which is a very complicated phenomenon. It could be useful to have a "particle-tracking" model to analyse this phenomenon. Moreover, in the Tien river system, the Ham Luong and Co Chien have the same tidal damping pattern. There is a sudden jump in the middle of the branches. For the Co Chien, this change also happens near the junction. Upstream and downstream of the jump, the tidal range is more or less the same, implying a balance between friction and convergence. However, as we discussed earlier, the tidal pattern in the Ham Luong and Co Chien is mostly determined by the dominant branch (i.e. the main Tien river). We suspect that downstream of the jump, the Ham Luong and Co Chien are driven by tidal influence from the mouths, while upstream of the jump, they are affected by the influence from the main Tien river.

We obtain very good calibration results on the Hau river for the hydrodynamic model. However, for the Tien river, which is a more complex system than the Hau, the calibration results are not satisfactory. There are several causes, which can be listed as follows: (i) insufficient and inadequate topographic data; and (ii) inadequate boundary conditions and hydraulic parameters.

On the other hand, the set of analytical equations has a limited applicability when applied to a complex system as the Tien river, as we can see in Figs. 5.5 and 5.7. The analytical equation is not good at analysing local interventions and disturbances (e.g. at several junction points), which can be seen clearly from the results of the hydrodynamic model. Moreover, the analytical equation assumes a completely natural, alluvial topography, which is not always the case in reality. The weakest equation seems to be the phase lag equation, which may have to be improved, particularly taking into account the effect of freshwater discharge and the effect of higher order tidal harmonics (overtides). Possibly, the bottom slope, which is often neglected but present in the Mekong, can also have an impact on the phase lag equation.

5.3.2 Scheldt estuary

The hydrodynamic model of the Scheldt estuary and its results

The Scheldt estuary (see Fig. 1.2) has a braided pattern of multi ebb and flood channels. Similar types of estuaries are found in Chesapeake Bay, USA (Ahnert, 1960); the Columbia river estuary, USA; the Pungue estuary, Mozambique; and in several small estuaries in the UK. Van Veen (1950) described the Scheldt estuary as an estuary with a regular system of ebb and flood channels. Jeuken (2000) identified two basic channel types in the Scheldt estuary: (i) main channels; and (ii) connecting channels. The main channels transport most of the tidal discharge, sediments and salinity. They are the largest channels in the system and they appear in two forms: a curved main ebb channel and a straight main flood channel. The connecting channels are the smaller channels that link either two large main channels or a large main channel with a small main channel. The transport function of the connecting channels is limited. Unlike the Mekong Delta, which contains a number of separate branches, in the Scheldt the ebb and flood channels interact closely through junctions and connecting channels.

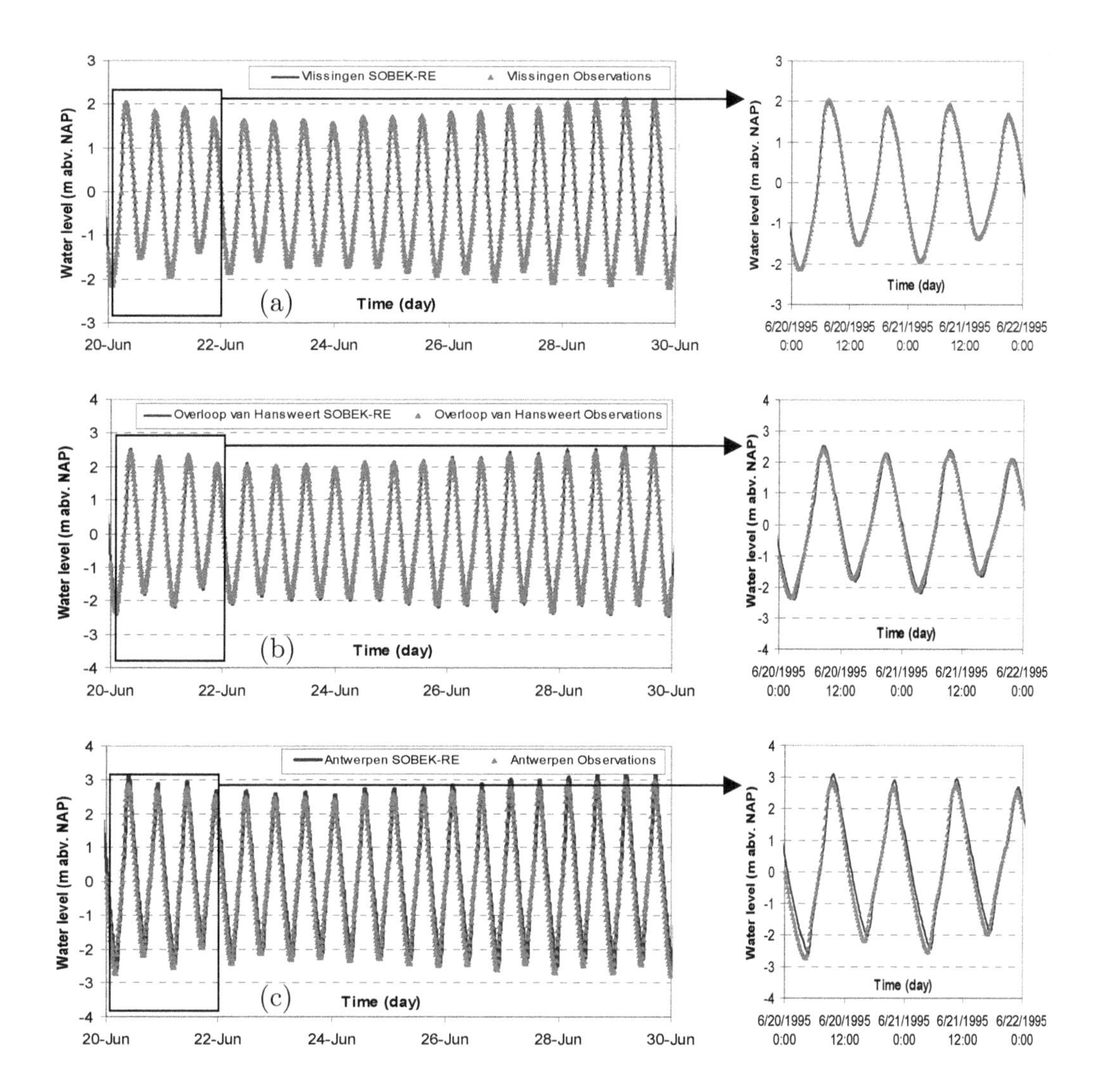

Figure 5.8 Model results for water level in the period from 20 to 30 June 1995 in the Scheldt estuary at locations: (a) Vlissingen, (b) Overloop van Hansweert, and (c) Antwerpen, showing modelled values (drawn lines) and observed values (triangles). NAP is the Dutch ordnance level.

The schematisation of the Scheldt in SOBEK-RE has been introduced earlier in Section 5.2.2. As an indication of model performance, in Figs. 5.8 and 5.9, we show the calibration results of the period 20-30 June, 1995 and the validation results of the period 20-30 June, 1998 at three stations along the Scheldt estuary: Vlissingen, Overloop van Hansweert (OvH) and Antwerpen, which are 27.5 km, 62.5 and 115 km from the open sea boundary (Vlakte van de Raan station – VR), respectively.

In Vlissingen the correspondence between the model and observations is near perfect. However, we notice deficiencies in modelling the correct amplitude and timing at other stations such as OvH and Antwerpen. For example, at OvH, timing of LW is slow; while at Antwerpen, HW and LW are high and timing of LW is slow. As we mentioned in Chapter 1 (Section 1.4), compared to the Mekong Delta, the Scheldt estuary has a better data set due to the dense station network as well as the smaller time-interval observations. This can be seen in Figs. 5.8 and 5.9.

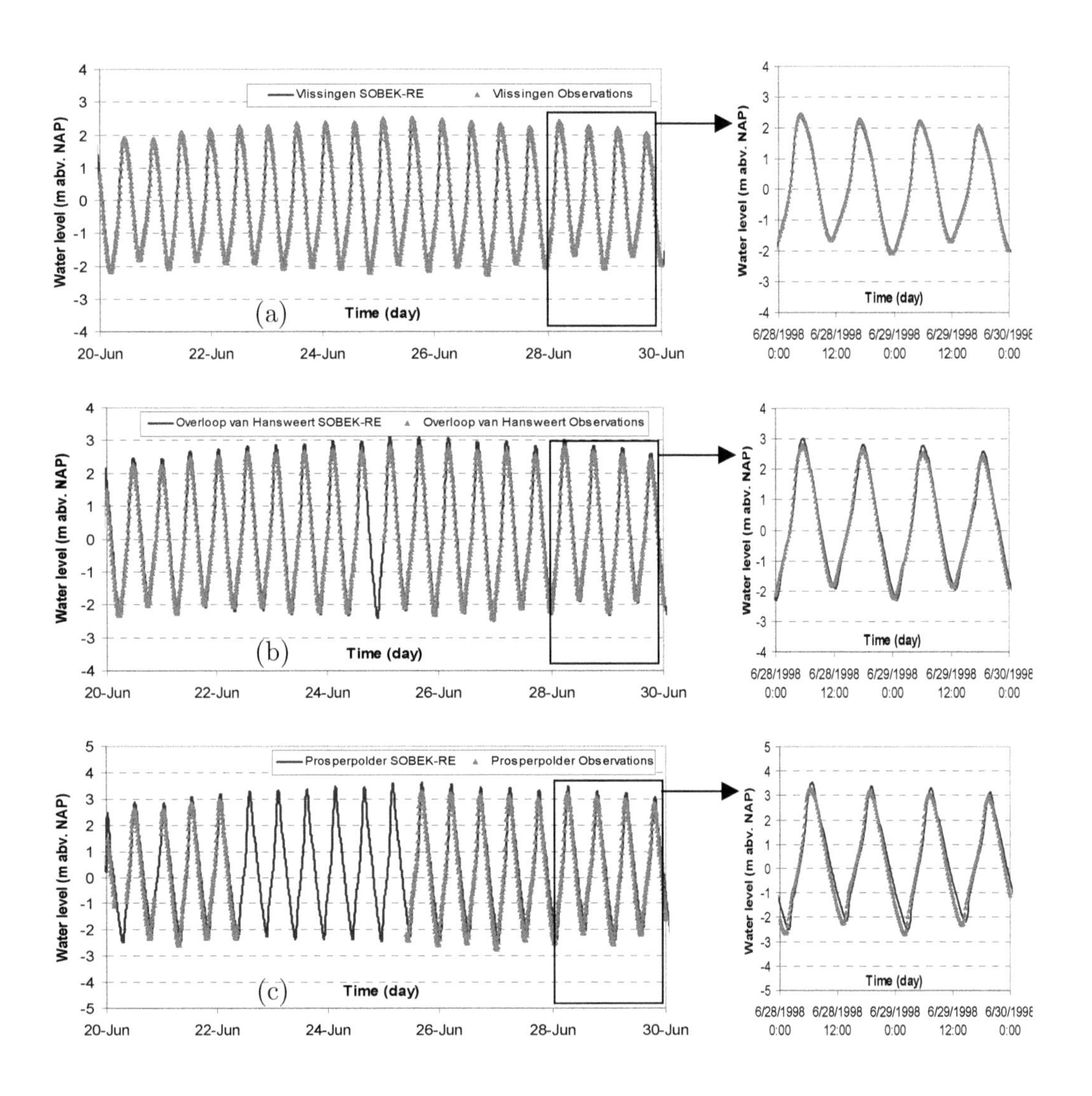

Figure 5.9 Model results for water level in the period from 20 to 30 June 1998 in the Scheldt estuary at locations: (a) Vlissingen, (b) Overloop van Hansweert, and (c) Antwerpen, showing modelled values (drawn lines) and observed values (triangles). NAP is the Dutch ordnance level.

Tidal damping and phase lag characteristics

Because the calibration on water levels is reasonable (see Figs. 5.8 and 5.9), we also obtain good results for tidal damping when we compare the hydrodynamic model with observations and the damping equation. The correspondence is good, especially in the amplified part of the estuary. In Fig. 5.10, we present the longitudinal development of the dimensionless tidal range y (i.e. the difference between HW and LW scaled on the tidal range at the downstream boundary), $y = H / H_0 = \eta / \eta_0$. However, it is noticed that there are some fluctuations in the hydrodynamic model results. This is probably an artefact of the schematisation, which uses discrete cross-sections and linear interpolation between them. The tidal damping patterns on the two different days are similar.

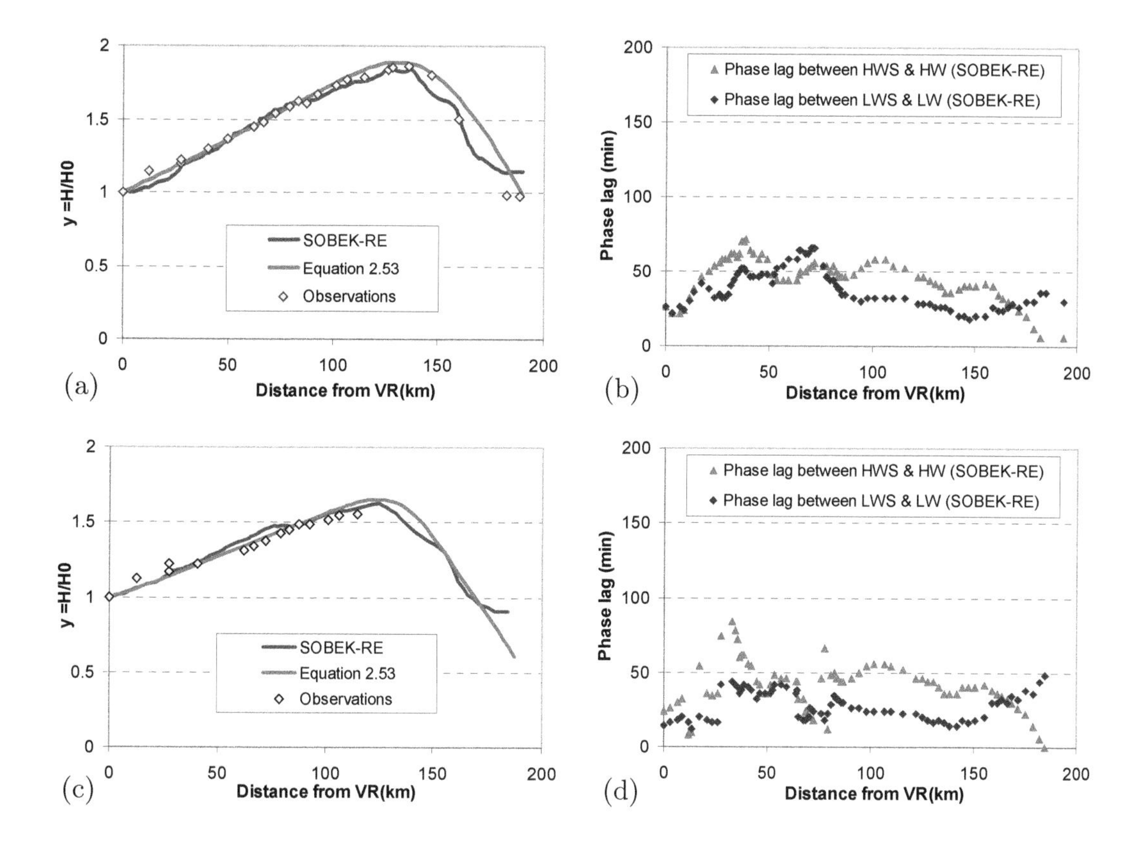

Figure 5.10 (a) Tidal damping and (b) phase lag in the Scheldt estuary on 21 June 1995 (neap tide); (c) Tidal damping and (d) phase lag (computed by SOBEK-RE) in the Scheldt estuary on 29 June 1998 (spring tide).

Unfortunately, reliable phase lag observations are not available from direct measurements. For the computed results in Figs. 5.10a and 5.10c, Savenije (2001) used a constant phase lag of 40 minutes. Looking at Figs. 5.10b and 5.10d, we observe the following: (i) at the sea boundary, the phase lag between HW-HWS and LW-LWS is similar because the tidal asymmetry is small; (ii) in the middle reach of the estuary, the phase lag between HW-HWS is higher than the phase lag between

LW-LWS due to the effect of overtides; (iii) near the upstream boundary, where the HW-HWS phase lag reduces to zero and the LW-LWS phase lag increases, the phase lag is influenced by the river discharge, which agrees with findings of Horrevoets *et al.* (2004). In general, the average value of the phase lag along the longitudinal axis of the estuary in the hydrodynamic model is about 30-50 minutes. We can see some sudden changes in the reach between 0 and 80 km, which are caused by connections between ebb and flood channels. Upstream of the 80 km point, there are no longer any connecting channels. We also can see the effect of connecting channels and junctions on the tidal damping and phase lag patterns, but it is not as strong as in the Mekong Delta.

Moreover, it is noticed that the phase lag values on 21 June 1995 (neap tide, $H_0 =$ 2.75 m) are larger than the values on 29 June 1998 (spring tide, $H_0 = 3.70$ m). This can be explained by the difference between neap tide and spring tide, just like we have seen in the Mekong Delta (see Section 5.3.1).

Tidal wave propagation

The characteristics of tidal wave propagation in the Scheldt estuary on 21 June 1995 and on 29 June 1998 are shown in Fig. 5.11. The wave celerity values at HW and LW are computed by using Eqs. 5.1 and 5.3. Because of the differences in depths and velocities, the celerity at HW and LW does not have the same propagation speed (Friedrichs and Aubrey, 1988; and Wang *et al.*, 2002). The behavior of the tidal wave propagation is similar to the findings of Dronkers (2005, pp. 252-253). The results obtained by the analytical equation are also presented in the same figure. Data used for the analytical equation have been taken from Savenije and Veling (2005) and Savenije (2005). We can see that the results of the hydrodynamic model do not fit the LW observations. Although the HW wave celerity agrees to the observations, the LW wave celerity of the model is much slower than the observation. We can also see in Figs. 5.8 and 5.9 that the computed times for LW lag behind the observations. There are three possible causes: (i) storage width and cross-sectional area in the model; (ii) roughness parameters; and (iii) inadequate account of dredging activities. In the following, we shall discuss these three possible causes.

Storage width and area

The hydrodynamic model of the Scheldt estuary only takes into account the main channel (without any storage width). However, the storage is implicitly taken into account due to a number of storage channels and the quasi-2D character of the network. Therefore, r_s in the model equals unity. Moreover, if r_s is larger than unity, then according to Eq. 2.45, the wave celerity would even be slower. Changing the storage does not resolve the problem.

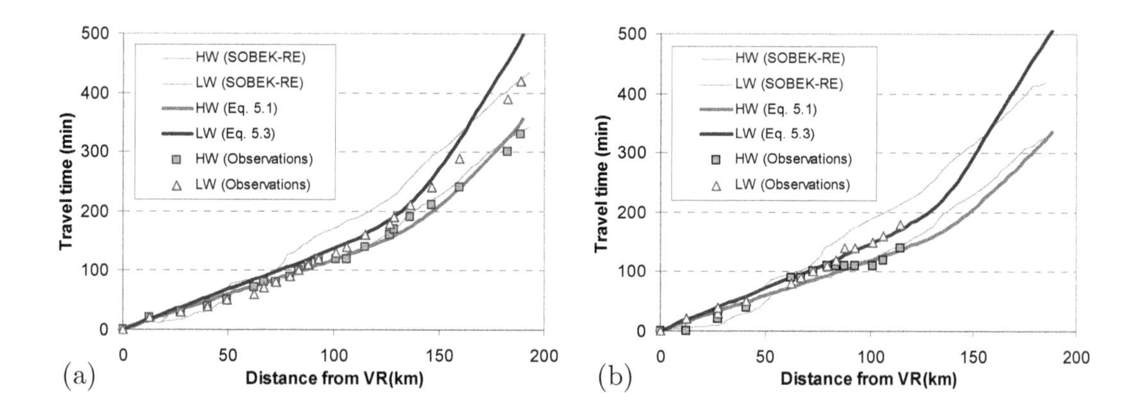

Figure 5.11 Tidal wave propagation in the Scheldt estuary: (a) on 21 June 1995 (neap tide) and (b) on 29 June 1998 (spring tide).

Roughness parameters

In the model, one of the most important hydraulic parameters is the friction presented by the Manning coefficient. The importance of the friction term has been recognized in Godin (1999) or Friedrichs and Aubrey (1988, 1994), especially for estuaries with damped tidal wave. Changing the Manning coefficient could possibly alter the characters of wave propagation, especially for damped estuaries where friction plays a more important role than convergence. However, for the Scheldt estuary, which is an amplified estuary, this is not the case. In Savenije and Veling (2005), it was shown that for an amplified wave, the model is not very sensitive to friction. Trials with the hydrodynamic model support this.

Effects of dredging

Wang *et al.* (2002, 2003) indicated that substantial annual dredging is required in the Scheldt estuary to maintain the shipping channel to the harbor of Antwerpen. The navigation channel was maintained at a minimum depth of 14.5 below NAP (Dutch ordnance level) until 1997. From 1997, the guarantied navigation depth was increased to 16.0 m below NAP. We notice in the model schematization that several locations in the navigation channel were much shallower than the guarantied navigation depth, especially at the shallow areas, where ebb and flood channels meet. This could be the reason why we have slower wave propagation. It is likely that the topography used in the model was surveyed before the annual dredging activities. Hence we have increased the depth at a limited number of points where the depth of the navigation channel was less than 14.5 m below NAP in order to see if this could explain the slower wave. However, we have not taken into account the topographical changes due to the effect of dumping and sand mining, which is the case for the Scheldt (Winterwerp *et al.*, 2001; and Kuijper *et al.*, 2004).

The simulated results show that if the navigation channel is maintained at -14.5 m, the tidal wave propagates faster (both HW and LW) due to the larger value of h.

Particularly, the change in the propagation of LW is noticeable (see Fig. 5.12) and the simulated results agree better with observations. However, it appears that if we would impose a deeper depth in the reach of 50-70 km and shallower depth in the reach of 70-100 km, we would even obtain a better fit with the observations. It appears that the wave propagation character is very sensitive to the topography even to local dredging of shallow crossover points. This confirms that in the Scheldt estuary, topography is the key factor determining the characteristics of the tidal wave propagation.

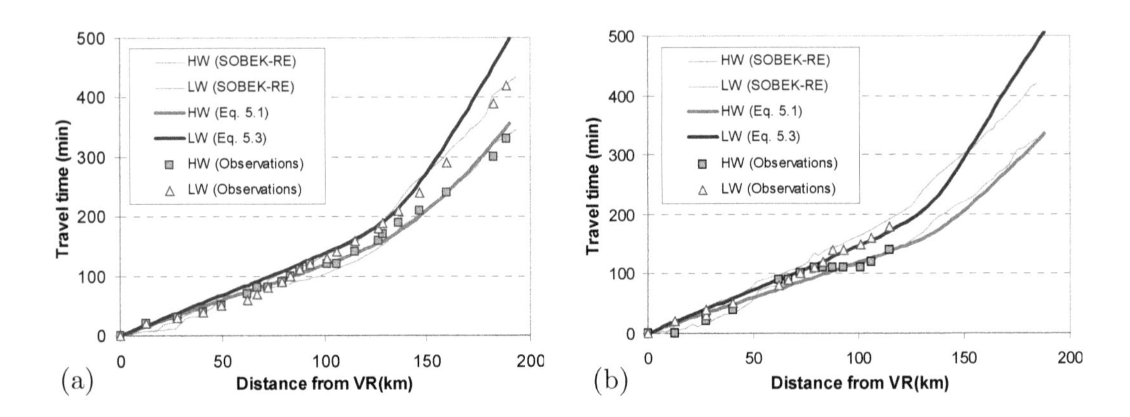

Figure 5.12 Tidal wave propagation in the Scheldt estuary (after adjusting the depth of the navigation channel): (a) on 21 June 1995 (neap tide) and (b) on 29 June 1998 (spring tide).

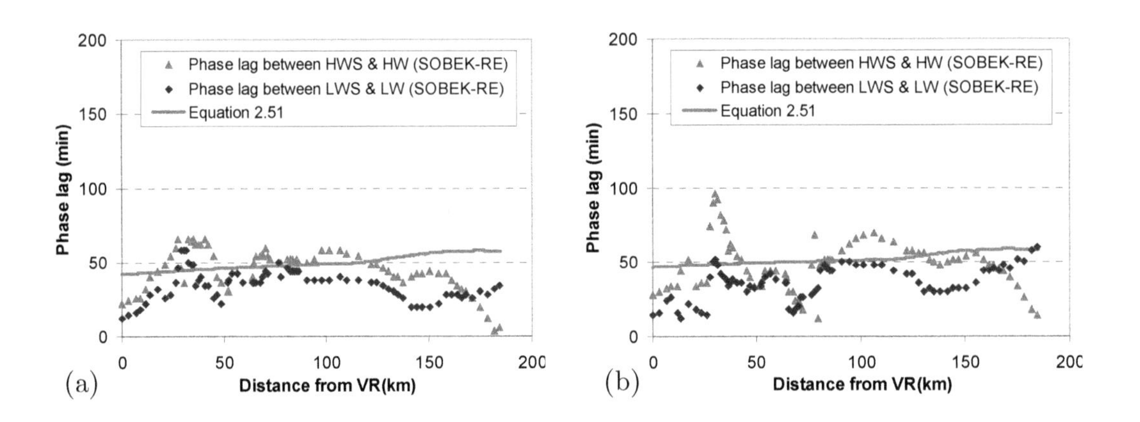

Figure 5.13 Phase lag in the Scheldt estuary (after adjusting the depth of the navigation channel): (a) on 21 June 1995 (neap tide) and (b) on 29 June 1998 (spring tide).

Does the local dredging affect the phase lag ε? There appear to be some changes in the phase lags due to the channel deepening (compare Figs. 5.10b and 5.10d and Fig. 5.13). Although the phase lag patterns are mostly similar before and after deepening the navigation channel, there is an overall adjustment in the order of 10 minutes. In general the phase lag values after deepening are larger than the original values. The increase of the phase lag can be explained as follows: Looking at Figs. 5.8c and 5.9c, which presents the situations before adjusting the navigation channel depth, the LW values at Antwerpen lag about 20-30 minutes behind the observations. After adjusting the channel depth, the LW values are more or less in phase with the observations. Assuming that the slack moments stay the same, then the ε values (after adjusting) are higher. The phase lag equation gives a good agreement with the hydrodynamic model until the 125km mark. Upstream of the 125km mark, probably due to the bottom slope and river discharge, similar to the case of the Mekong branches, the phase lag equation does not give a good agreement.

Discussion

Although producing reasonable results for the water level variation, the hydrodynamic model has difficulty to correctly represent the wave propagation. The main reason is that the schematization in the hydrodynamic model does not represent the real topography. In addition, it uses linear interpolation between cross-sections. Another weak point of the hydrodynamic model is that in order to evaluate the effect of dredging, the whole topography needs to be changed and the model needs to be re-run and tuned before effects can be seen. In the analytical equation, these effects can be seen directly. In addition, it is possible that the solution algorithm of the hydrodynamic model does not fully account for the non-linear terms in the momentum equation resulting in not fully correct asymmetry of the tide.

The analytical equations can be useful to facilitate the calibration of the hydrodynamic model. Especially for rivers and estuaries where there are not many observations available, the analytical equations form a good tool to interpolate between observations and to check the reliability of a hydrodynamic model. It is recommended that calibration should not only focus on water level but also on velocity (e.g. occurrence of slack moments) and wave propagation (timing of HW, LW). However, the analytical equation is not good at analyzing local interventions and disturbances; and it assumes a completely natural, alluvial topography, which is not always the case in reality.

5.3.3 Comparison with analytical equations

It has been demonstrated in the previous sections that the analytical equations, despite their simplicity, perform almost identical to the hydrodynamic models for the cases of the Mekong Delta and the Scheldt estuary. Only upstream of these estuaries due to effects of freshwater discharge and bottom slope, the phase lag equation gives inaccurate results compared to the hydrodynamic model results. This could be due to disadvantages of both the analytical equation and the hydrodynamic models. Nevertheless, we can see that they indeed can present a reliable picture of the study areas.

Savenije *et al.* (2007) showed that classification of estuaries should be based on two parameters, the estuary shape number γ and the friction scale χ. Two types of estuaries can be distinguished: strongly convergent and weakly convergent estuaries. They have different behavior. For large values of γ (i.e. strongly convergent estuaries), the behavior no longer depends on the friction scale χ. For smaller values of γ (i.e. weakly convergent estuaries), it strongly depends on χ. We shall proceed to see how the study areas fit into this classification.

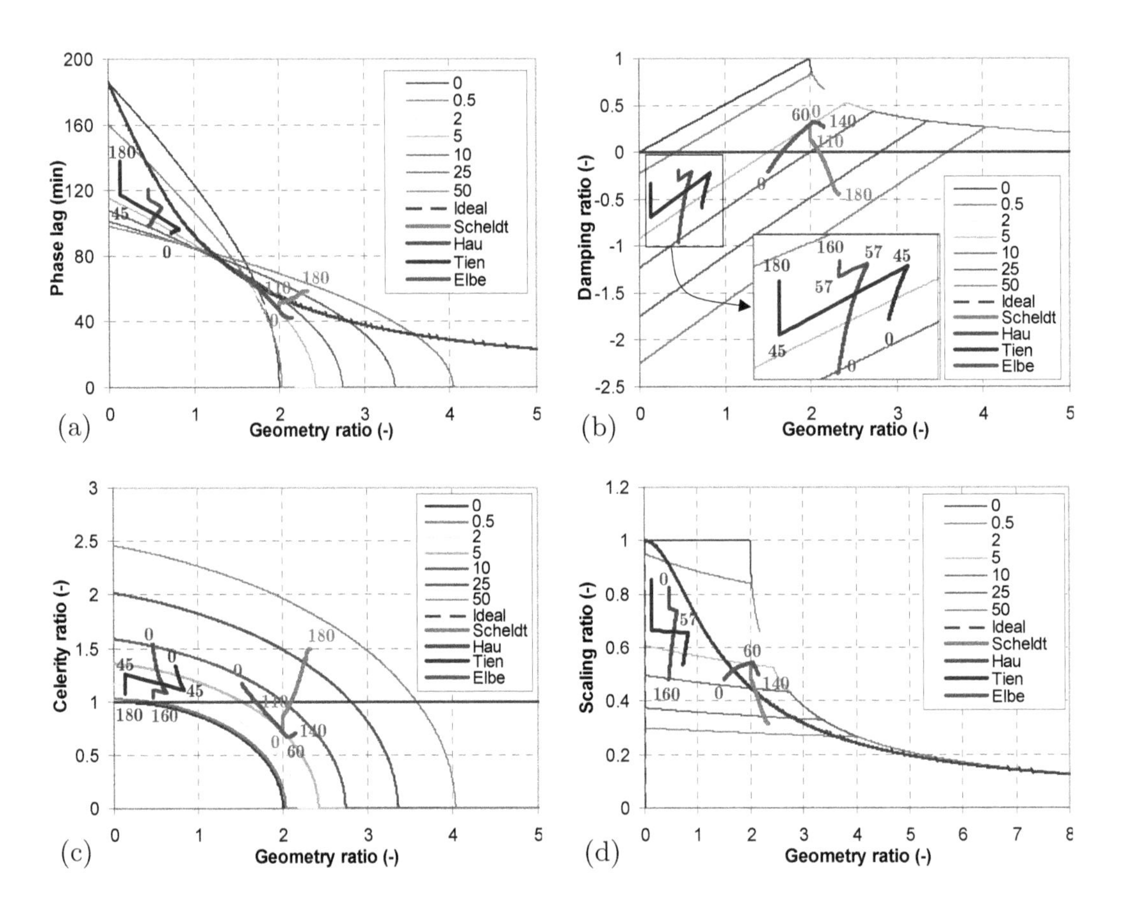

Figure 5.14 Positioning of the Schelde (red), Hau (brown) and Tien (dark blue) estuaries in: (a) phase lag diagram; (b) damping number diagram; (c) celerity number diagram; and (d) velocity number diagram. The blue line represents the "ideal" estuary and next to the segments, the distance in km is written indicating the length over which a segment is representative (Figs. 5.14b and c). The data of the Elbe, which is considered as an ideal estuary, have been taken from Haas (2007).

Applying the set of four equations (i.e. Eqs. 2.55 - 2.58 in Section 2.5.2, Chapter 2), we are able to estimate the values of the three main factors for describing the tidal wave characteristics in the Mekong Delta branches and the Scheldt estuary. Figure 5.14 presents the computed results. Looking at Fig. 5.14, the Hau and Tien estuaries behave similarly. They are close relatives. This can be perfectly understood given the same hydrological and topographical conditions within the Mekong Delta. They are mostly riverine in character. These estuaries have a small estuary shape number, therefore have a large phase lag and a damped tidal wave. In contrast, the Scheldt estuary is a more marine estuary with a large estuary shape number, a smaller phase lag and an amplified wave. We can see that these estuaries are close to the "ideal" estuary. This implies that natural-alluvial estuaries tend to reach an ideal state of constant energy dissipation. This is similar to what Rodriguez-Iturbe and Rinaldo (2001) observed in river basins.

5.4 CONCLUSIONS

This chapter presents an analysis on the characteristics of the tidal wave in the branches of the Mekong Delta and the Scheldt estuary on the basis of the set of analytical equations, the hydrodynamic models and observations.

The Mekong Delta branches are mostly riverine in character. They have a small estuary shape number, therefore they have a large phase lag and a damped tidal wave. The damped tidal wave in the Mekong Delta estuarine branches moves slower than indicated by the classical celerity equation. The junctions between two branches have a significant influence on the pattern of the tidal range, phase lag and tidal wave propagation. It is concluded that in a complex system containing a number of branches, there is a dominant branch, which represents the characteristics of the system and it forces other branches to adapt and to adjust. For the case of the Tien river, the My Tho river is the dominant and the longest branch. For the less complex system (i.e. the Hau river, which contains only two paired branches with the same length), the analytical equations agree very well with observations and with the hydrodynamic model. However, for the Tien river, which is a complex system containing three big branches and five sub-branches, the agreement between these three sources is less convincing. Therefore, it is concluded that the set of analytical equations is not yet applicable for a complex system like the Tien river; and the hydrodynamic model should be further calibrated when detailed data on topography and hydrology is available and the calibration process should be done with special care to the junctions of the branches.

The Scheldt estuary is a marine estuary with a large estuary shape number, a small phase lag and an amplified wave. It appears that the tidal wave propagation in the Scheldt estuary is faster than suggested by the classical celerity equation due to the amplification of the tidal range. It is concluded that the tidal wave celerity depends mainly on the topography.

Finally, the comparison between these three sources (i.e. the set of analytical equations, the hydrodynamic models and observations) leads to the advice that the hydrodynamic model should be calibrated not only on water level and velocity amplitudes, but also on the occurrence of slack and on wave propagation characteristics. On the other hand, the analytical equations, despite their simplicity, are able to describe the tidal dynamics in the Mekong Delta and the Scheldt estuary, which are both complex multi-channel systems. However, it is recommended to further improve the phase lag equation by taking into account effects of the freshwater discharge and the bottom slope.

Chapter 6

MIXING IN ESTUARIES WITH AN EBB-FLOOD CHANNEL SYSTEM

Abstract

Tidal pumping caused by residual horizontal circulation is an important but ill-understood mechanism producing longitudinal salt dispersion in well-mixed estuaries. There are two types of residual circulation causing tidal pumping: (i) interaction of the tidal flow with a pronounced flood-ebb channel system; and (ii) interaction of the tidal flow with the irregular bathymetry. Residual ebb-flood channel circulation is an important large-scale mixing mechanism for salinity intrusion, as shown in the Western Scheldt in the Netherlands, which is a well-mixed estuary with a distinct ebb-flood channel system. This chapter provides a new simplified conceptual model and a new analytical equation for this type of mixing. Firstly, using a fully three-dimensional hydrodynamic model as a "virtual laboratory" and employing a decomposition method, the characteristics of the residual ebb-flood channel circulation in the Western Scheldt have been analysed. The analysis shows that over one tidal cycle, there is an upstream residual salt transport in the flood channels and a downstream residual transport in the ebb channels. Secondly, a conceptual model and an analytical equation determining the dispersion coefficient have been developed, which take into account relevant parameters for tidal pumping, such as the tidal pumping efficiency and the length of the ebb-flood channel loops. Subsequently, the newly developed equation has been compared to the results of the "virtual laboratory" and a steady-state salt intrusion model. The comparison confirms an agreement between the newly developed equation and the existing models in determining the residual transport and the tidal pumping dispersion coefficient. Finally, the equation has been applied to observations in the Western Scheldt and the Pungue estuary. The application yields good results in determining the longitudinal dispersion compared to dispersion values obtained from the salt budget.

Parts of this chapter were:

Nguyen, A.D., Savenije, H.H.G., van der Wegen, M., and Roelvink, J.A., 2007. New analytical equation determining the dispersion coefficient in estuaries with a distinct ebb-flood channel system. Accepted for publication in Estuarine, Coastal and Shelf Science.

Nguyen, A.D., Savenije, H.H.G., van der Wegen, M., and Roelvink, J.A., 2007. Mixing in estuaries with a distinct ebb-flood channel. In D. Boyer, O. Alexandrova (Eds.), Proceedings of the Fifth International Symposium on Environmental Hydraulics, Tempe, Arizona, USA, 6pp.

6.1 INTRODUCTION

The tidal pumping mechanism, which is caused by residual horizontal circulation, is an important but ill-understood part of tidal circulation producing longitudinal salt dispersion in estuaries. Fischer *et al.* (1979) defined "tidal pumping" as the energy available in the tide that drives steady circulations similar to what would happen if pumps and pipes were installed to move water about in circuits. There are two types of residual circulation that cause tidal pumping: (i) interaction of the tidal flow with a pronounced flood-ebb channel system; and (ii) interaction of the tidal flow with the irregular bathymetry. It is expected that the residual ebb-flood channel circulation is an important large-scale mixing mechanism for moving pollutants and transporting salinity upstream against a mean outflow of freshwater in estuaries with a distinct ebb-flood channel system, such as the Western Scheldt in the Netherlands (see Fig. 1.2). The similar type of estuaries can be found in the Chesapeake Bay, USA (Ahnert, 1960); the Columbia river estuary in USA; the Pungue estuary in Mozambique as well as several small estuaries in UK. Van Veen *et al.* (2005) described the Western Scheldt as an estuary with a regular system of ebb and flood channels. Jeuken (2000) identified two basic channel types in the Western Scheldt estuary: (i) main channels and (ii) connecting channels. The main channels transport most of the tidal discharge, sediments and salinity. They are the largest channels in the system and they appear in two forms: a curved main ebb channel and a straight main flood channel. The connecting channels are the smaller channels that either link two large main channels or a large main channel with a small main channel, and their transport function is limited.

Savenije (1993b) found that gravitational (density-driven) circulation is the main mixing mechanism if salinity gradients are large (as occurs in the central part of estuaries). The river flow plays an important role to drive the gravitational circulation, which is accompanied by both vertical and lateral salinity gradients. In wide estuaries, lateral stratification generally makes the largest contribution to density-driven mixing (Fischer *et al.*, 1979). However, in the seaward part of estuaries, where the density difference is small, the interaction between tide and geometry, which generates residual (horizontal) circulation, is the main mixing mechanism. McCarthy (1993) presented an analysis on residual circulation generated by the combined effect of tide and gravitational circulation in an estuary with exponentially varying width. Although his study did not consider the important transport mechanism by ebb-flood channel interaction, it clearly demonstrated that tide-driven mixing is dominant in the downstream part of estuaries, while density-driven mixing is dominant in the upstream part of estuaries. Density-driven mixing is a function of the salinity gradient, whereas tide-driven mixing is a function of the salinity and the width.

Based on comprehensive reviews of Fischer *et al.* (1979), Zimmerman (1986), and Geyer and Signell (1992), it can be seen that many authors have tried several different methods to quantify tidal-driven dispersion. Several authors used a decomposition method (e.g. Fischer, 1976; Lewis, 1979; Uncles and Jordan, 1979; Uncles *et al.*, 1985; Pino Q *et al.*, 1994; Dyer, 1997; or Sylaios and Boxall, 1998). A number of authors proposed to determine the effective longitudinal tidal-driven dispersion as a function of mixing length and velocity. Examples of this approach can be found in Arons and Stommel (1951) and Zimmerman (1976). Arons and Stommel proposed $D = k_A U_0 l_0$ (where D, k_A, U_0, l_0 are the dispersion coefficient, a constant, velocity scale and mixing length scale, respectively). k_A is understood to be a constant for one estuary. However, because the values of k_A widely varied, it resulted in a very large range of computed values. Zimmerman (1976) developed the "tidal random walk" theory and model, which considers the Lagrangean motions in estuaries resulting from the purely advective effects of tidal and residual currents and takes into account pronounced horizontal residual circulations generated by tide-topography interactions. Due to the presence of residual eddies, the velocity field cannot be considered uniform and in fact it varies considerably over distances of the order of the tidal excursion. Zimmerman (1976, 1981) derived an equation for the longitudinal dispersion coefficient, which is formally the same as the equation of Arons and Stommel. The main innovation in the equation of Zimmerman is that k_A has a well-defined physical meaning, being a function of the dimensionless parameters reflecting mixing length and tidal velocity. However, to obtain reasonable values for k_A, good quality and detailed observations are needed. Zimmerman applied his theory to a lagoon system (i.e. the Dutch Wadden Sea) with not as well-defined channels as occur in alluvial estuaries studied herein. The fact that the topography of alluvial estuaries obeys geometrical laws creates an opportunity to expand this theory and make it more predictive.

Furthermore, de Swart *et al.* (1997) applied the "tidal random walk" model to the Ems estuary and obtained good results, at least qualitatively, in estimation of the dispersion coefficients. However, they recommended that although the simple random walk model can be used to get upper bounds for the dispersion coefficients, simulations using a detailed 3D-numerical model are necessary in order to obtain lower bounds for the values of dispersion coefficients. This well agrees with assessments in Geyer and Signell (1992), which stated that numerical models provide a means of isolating the influence of tidal advection from these other processes, and they can simulate the nonlinear effects that produce residual circulations.

The aims of this chapter are to investigate the residual circulation caused by the interaction of the tidal flow with the flood-ebb channel system in the Western Scheldt and to study the effect of tidal pumping caused by the residual circulation on the salinity distribution. This chapter provides a new conceptual model and a new analytical equation for this type of tidal-driven mixing. Firstly, we use a fully three-dimensional hydrodynamic model as a "virtual laboratory" and employ a decomposition method to characterize the residual ebb-flood channel circulation in the Western Scheldt. Secondly, a conceptual model and an analytical equation determining the dispersion coefficient are developed, which take into account several important parameters of the tidal pumping mechanism, such as the tidal pumping efficiency and the loop length. Subsequently, the newly developed equation is compared to the results of the "virtual laboratory" and a salt intrusion model. Finally, the equation is applied to observations in the Western Scheldt and the Pungue estuary. The application yields good results in determination of the longitudinal dispersion compared to dispersion values obtained from the salt budget and hence it is demonstrated that the newly developed equation can be applied to a real estuary.

6.2 METHODOLOGY

6.2.1 "Virtual laboratory" residual salt transport

In order to analyze the characteristics of the residual ebb-flood channel circulation, a considerable set of data on residual flow components is required. The collection of such information would require a major operation, consisting of a dense network of monitoring points over a considerable period. Moreover, substantial errors on measurements may result (Lane *et al.*, 1997). In practice, it is not feasible to obtain these data from the field. Therefore, in this chapter, a fully three dimensional hydrodynamic model (DELFT3D) has been used as a "virtual laboratory" of the Western Scheldt estuary to provide data on hydrodynamics and salinity. For details on the operation of the hydrodynamic model and the key features of the formulations used, reference is made to Lesser *et al.* (2004).

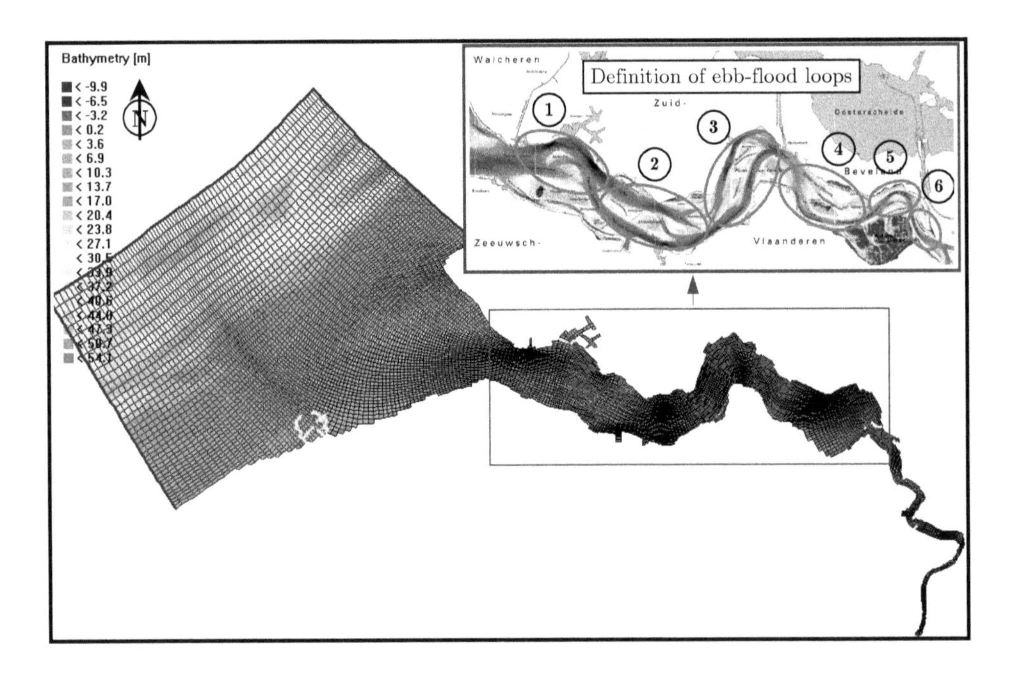

Figure 6.1 The Western Scheldt estuary schematization in DELFT3D (not to scale).

The plan-view of the Western Scheldt estuary schematization in the hydrodynamic model is shown in Fig. 6.1. Some relevant information on the schematization of the hydrodynamic model is summarized as follows:

(i) Configuration: In total the grid includes approximate 12,500 active points, grid cells sizes vary from 800 m × 800 m at the open sea to 150 m × 50 m within the estuary and the number of layers is 10. The bathymetry data used are of 1998 and the simulation period is three months.

(ii) Boundary conditions: At the downstream tidal boundary we use a tidal amplitude of 1.75 m (i.e. tidal range of 3.5 m, which is a typical value for the

Western Scheldt estuary), using the gradient boundary conditions (a so-called Neumann boundary condition, Roelvink and Walstra, 2004) with the tidal period of 12.25 hours (i.e. semi-diurnal tide). Because the upstream boundary conditions are still located in the tidal zone, we imposed a harmonic function for the discharge in order to have a realistic discharge profile. This harmonic function is based on a 1-D model and has an amplitude of 4,200 m^3/s and an average discharge of 120 m^3/s, which is a typical value for the studied estuary.

The hydrodynamic model is derived from an existing hydrodynamic model, denoted as KUSTZUID, which was set-up, calibrated and verified by Rijkswaterstaat (Dutch Ministry of Transport and Public Works). As the original KUSTZUID model was already thoroughly calibrated by Rijkswaterstaat no extensive calibration appeared to be necessary (Kuijper *et al.*, 2004). The hydrodynamic regime (i.e. water level, velocity and discharge) of the model was calibrated and validated on the basis of data in 2000 and 1978, respectively. The salinity simulation was calibrated based on available data in September 2002. Generally, computed water levels, discharges and salinities are in good agreement with observations, while velocity measurements taken in the center of a channel are reasonably well reproduced by the model.

In order to obtain values for the residual salt transport in the "virtual laboratory", the decomposition method of Uncles *et al.* (1985) and Pino Q *et al.* (1994) has been employed. A brief summary on the method has been presented in Chapter 2, Section 2.3.2. The residual transport of salt in the "virtual laboratory" has been computed for separate ebb and flood channels as well as for the entire estuarine channel. As indicated in Van Veen *et al.* (2005) and Jeuken (2000, pp. 31-36), the flood channel is wide and shallow, and the ebb channel is narrow and deep. The separation of the ebb and flood channels is therefore based on the highest point of the shallow area between two channels. This separation point can be recognized from the topographical map. The residual salt transport contains three main components: the total residual transport, tidal pumping transport and gravitational transport. For reasons of simplicity, in the remaining sections, the residual salt transport and its components computed by applying the decomposition method will be called the "virtual laboratory" results.

6.2.2 Mixing in a hypothetical ebb-flood channel estuary

In this section, a schematized numerical model is introduced. The model is developed to analyze effects of residual circulation on salinity distribution in a hypothetical estuary with a distinct ebb-flood channel system and to investigate the relation between the longitudinal salt transport and relevant parameters.

Model equation

It is assumed that the movements of salt in the hypothetical estuary can be presented by the advection-dispersion equation:

$$\frac{\partial S}{\partial t}+\overline{u}\frac{\partial S}{\partial x}=\frac{1}{A}\frac{\partial}{\partial x}AD\frac{\partial S}{\partial x} \tag{6.1}$$

where D (L^2T^{-1}) is the effective longitudinal dispersion coefficient.

Equation 6.1 is the simplified version of Eq. 2.26 due to the absence of r_s and R_S. For a steady state, at HWS, LWS or TA conditions, we have $u_f A = Q_f$ (which is constant with x). Integration of Eq. 6.1 over x, under the boundary condition that $dS/dx=0$ at $S=S_f$ results in:

$$\left(\langle\overline{u}\rangle A\right)S = AD\frac{\partial S}{\partial x} \tag{6.2}$$

in which $\left(\langle\overline{u}\rangle A\right) = Q_f$.

In a Eulerian reference frame, the cross-sectional average velocity in the system can be represented by (Savenije, 2005):

$$\overline{u} = \upsilon\sin(\xi) \tag{6.3}$$

where υ (LT^{-1}) is the tidal velocity amplitude, and:

$$\xi = \omega t - \frac{\omega(x - x_0 - X)}{c}. \tag{6.4}$$

where ξ (-) is the dimensionless argument; ω (T^{-1}) is the angular velocity, $\omega = 2\pi/T$; and X (L) is the distance traveled by a water particle, $X = (\upsilon/\omega)(1-\cos(\omega t))$. If we move with the water particle ($x = x_0 + X$), then the Lagrangean equation is obtained with $\xi = \omega t$ and $\overline{u} = \upsilon\sin(\omega t)$.

In this schematized conceptual model, in order to analyze the effect of the large-scale residual ebb-flood channel circulation, we assume that mixing only takes place in the nodes and not in the ebb and flood branches themselves and the cross-section area is constant, Eq. 6.1 then becomes:

$$\frac{\partial S}{\partial t}+\overline{u}\frac{\partial S}{\partial x}=D\frac{\partial^2 S}{\partial x^2} \tag{6.5}$$

Equation 6.5 is a Lagrangean equation, which is solved by the QUICKEST scheme (Abbott and Basco, 1989), which has a good stability and accuracy. The QUICKEST scheme reads:

$$S_j^{t+1} = \left[\gamma(1-C_R)-\frac{C_R}{6}(C_R^2-3C_R+2)\right]S_{j+1}^t - \left[\gamma(2-3C_R)-\frac{C_R}{6}(C_R^2-2C_R-1)\right]S_j^t$$
$$+\left[\gamma(1-3C_R)-\frac{C_R}{6}(C_R^2-C_R-2)\right]S_{j-1}^t + \left[\gamma(C_R)+\frac{C_R}{6}(C_R^2-1)\right]S_{j-2}^t + S_j^t \tag{6.6}$$

in which $\gamma = \dfrac{D\Delta t}{\Delta x^2}$ and $C_R = \dfrac{\overline{u}\Delta t}{\Delta x}$; j and t indicate space and time steps, respectively.

Model schematization

The movements of water particles in the hypothetical estuary with a distinct flood-ebb channel system are modeled in MATLAB code. The model's schematization consists of six loops (see Fig. 6.2), similar to the Western Scheldt estuary.

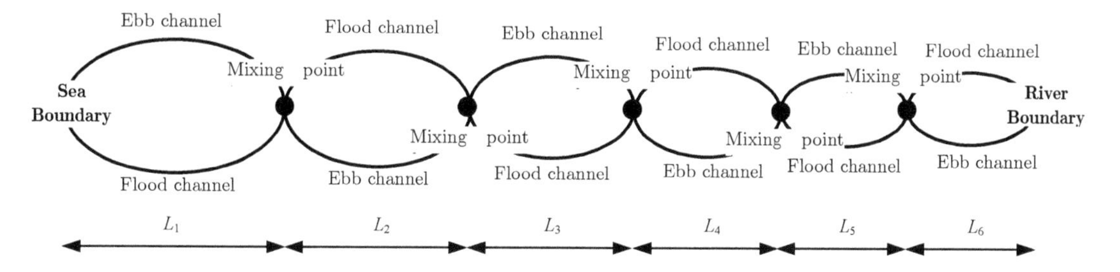

Note: L_i (i =1÷6) is the average loop length

Figure 6.2 Schematisation of the hypothetical estuary.

Model input and output

The model requires several inputs: (i) the tidal velocity amplitudes in flood and ebb channels and the lengths of the six loops; (ii) water and salinity boundaries at the sea and the river; and (iii) initial conditions. The estuary is modeled as a mixing zone between the sea and the river.

The model produces several outputs: (i) salinity at the six mixing points as a function of time (ii) the salinity distribution in each (flood or ebb) channel as a function of time.

To simplify the model, certain assumptions have been made:

(i) The estuary is well-mixed.

(ii) At the sea boundary, the velocity obeys Eq. 6.3. At the river boundary, a constant river velocity is taken.

(iii) The mixing happens only at the crossover points, while in reality there are some exchanges between channels and there is mixing within a loop. The mixing at the crossover points is proportional to the transport in the branches $S=\left(S_f v_f + S_e v_e\right)/\left(v_f + v_e\right)$, where S_f and v_f are the salinity and velocity in the flood channels, and S_e and v_e the salinity and velocity in the ebb channels. In addition, it is assumed that the cross-sectional area of the ebb channel equals to that of the flood channel.

Model results

The tidal pumping efficiency e_p is defined as:

$$e_p = \frac{\Delta v}{\tilde{v}_f} \tag{6.7}$$

in which Δv (LT^{-1}) is the difference between the average flood and ebb flow velocities in the flood channel. $\tilde{v}_f$ (LT^{-1}) is the cross-sectional average velocity during the flood tide.

In Fig. 6.3, the simulation results are presented. In the most downstream loop, pure seawater fills the flood channel, while somewhat fresher water flows down the ebb channel. In the second loop the same phenomenon happens, resulting in a decreasing salt intrusion. If the loop length and the tidal pumping efficiency are increased, the longitudinal dispersion increases proportionally. A longer loop length causes further salt intrusion because salt water can intrude further inland during a tidal period. In addition, a higher tidal pumping efficiency causes a higher salinity because more salt water can be pumped into the estuary within a tidal period. The salt intrusion follows the dome shape, and the longer is the loop length, the stronger is the dome shape.

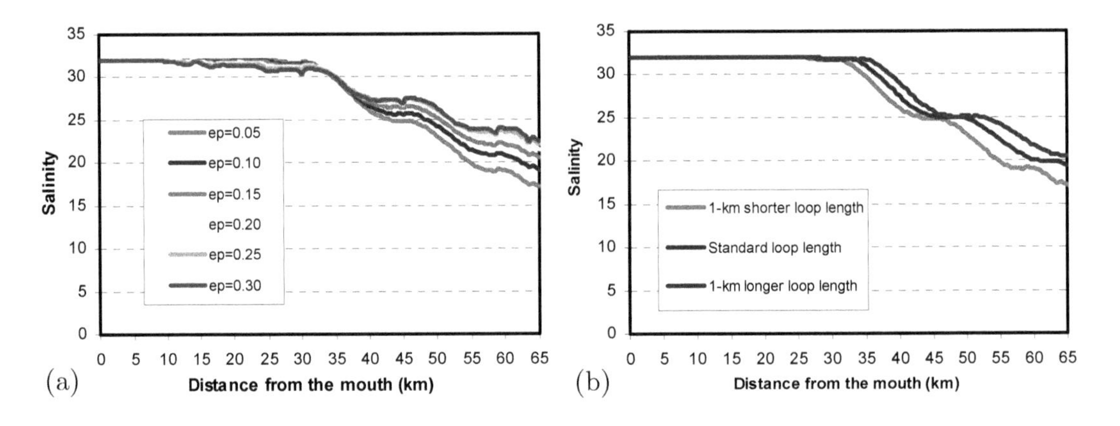

Figure 6.3 Lagrangean model results for longitudinal salinity distribution at HWS, considering: (a) different pumping efficiencies e_p; and (b) different ebb-flood loop length L_{ef}.

The above model is simple and the results are, of course, dependent on the assumptions made. However, the results reveal that indeed there is a direct relation between the longitudinal salt transport and the ebb-flood loop length, as well as the tidal pumping efficiency. Therefore, the loop length and the tidal pumping efficiency will be considered as the two most important parameters for further analysis.

6.2.3 Conceptual model and analytical equation for dispersion by residual ebb-flood channel circulation

In this section, we present the development of a conceptual model and an analytical equation for determining the dispersion coefficient in estuaries with a distinct ebb-flood channel system.

Analytical relation between loop length and width in ebb-flood channel estuaries

We analyze three estuaries, which can be considered to have an ebb-flood channel system. They are: the Western Scheldt in the Netherlands, the Columbia in USA and the Pungue in Mozambique. It is remarked that the Columbia may not be an ideal example for this type of estuaries since the distinction between the ebb and flood channels is not clearly visible, especially at the estuary mouth.

An ebb-flood channel loop is defined on the basis of its topographical characteristics. A loop is determined by a reach between two points (shallow areas) where the flood and ebb channels meet (see Fig. 6.1 for the case of the Western Scheldt). This definition is similar to the "morphological cells" defined by Winterwerp *et al.* (2001) for the Western Scheldt. The loop length can be considered as the "effective" horizontal wavelength, which is analogous to the definition of Zimmerman (1976) and it is an important parameter for determining residual dispersion (Zimmerman, 1981).

The shape of these three estuaries corresponds very well with the exponential functions that follow from the concept of ideal estuaries (see Fig. 6.4 and Table 6.1). The exponential functions of the area and the width have already been introduced in Chapter 3.

Table 6.1 Estuarine characteristics of three studied estuaries

Estuary	A_0 (m^2)	B_0 (m)	$\bar{h}$ (m)	a (km)	b (km)
Scheldt	100,000	10,000	10.0	28	28
Pungue	28,000	6,500	4.3	20	20
Columbia	40,000	7,000	5.7	45	45

Note: The values for the width, depth and cross-sectional area were measured at Mean Sea Level.

It appears that there is a certain width at which separation between ebb and flood channels no longer occurs. For the three considered estuaries, this width is in the order of 1.5 km. We can see in Fig. 6.4 that the last visible loop is constrained by a width of about 1.5 km. The topographical data suggest that for an ebb-flood channel system the loop length (i.e. the average length of the flood channel and ebb channel in the considered loop) and the loop width (i.e. the average width of the considered loop) are related.

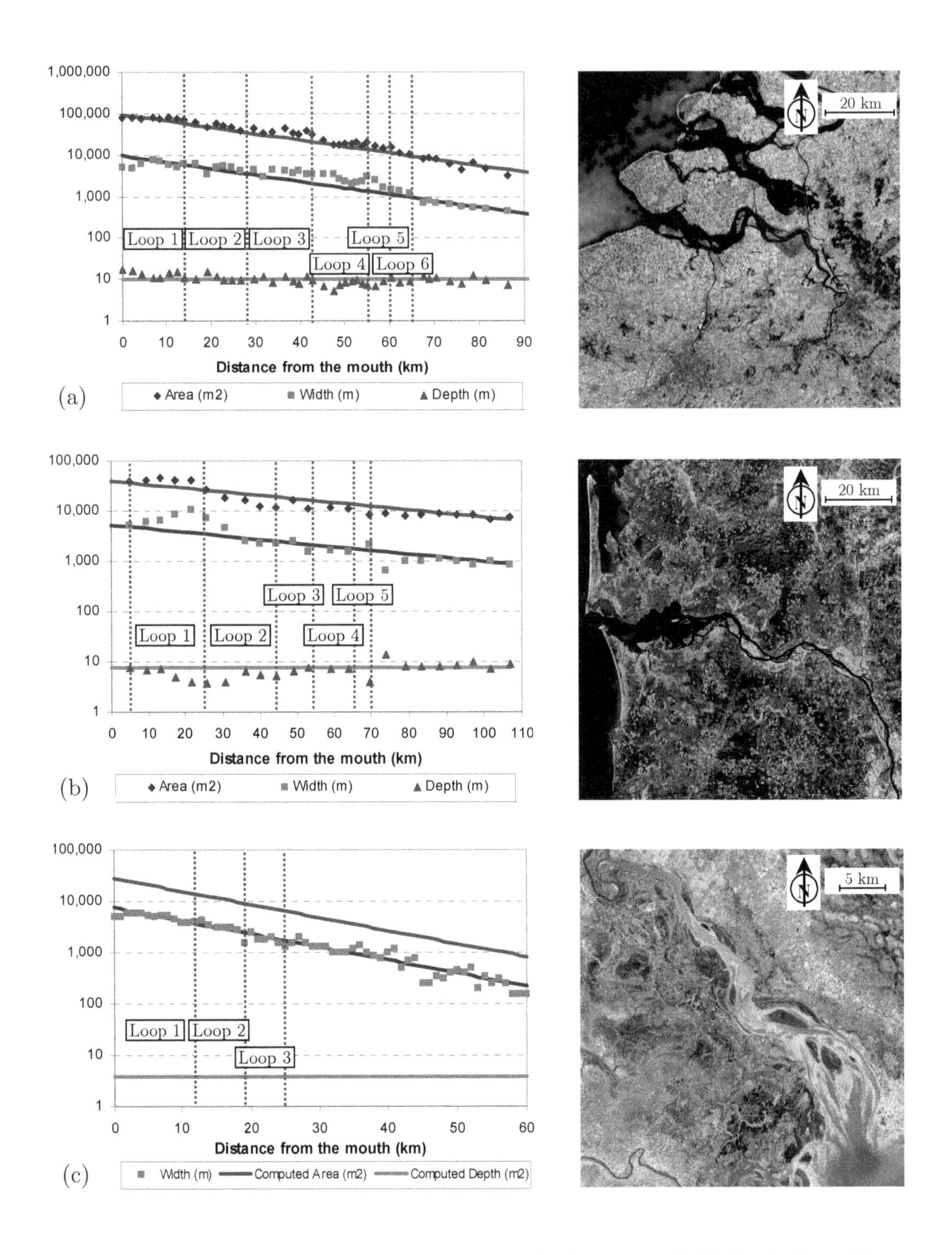

Figure 6.4 Shape of considered estuaries, showing the locations of ebb-flood channel loops: (a) Western Scheldt, The Netherlands (b) Columbia, USA; and (c) Pungue, Mozambique. The location maps have been taken from Google Earth.

From the available data of these three estuaries, it is possible to derive an empirical relation between loop length and loop width. Figure 6.5 shows the parity diagram for the following relation:

$$\frac{L_{ef}}{b} = \alpha_L \left(1 - \frac{B_L}{B_0 - B_L}\left(\frac{B_0}{B} - 1\right)\right) \tag{6.8}$$

where L_{ef} (L) is the loop length; α_L (-) is a scaling coefficient that is approximately equal to 0.5; B_0 (L) is the width at the estuary's mouth. B_L (L) is the channel width at the point where there is no more ebb-flood channel separation, b (L) is the width convergence length and B (L) is the averaged loop width.

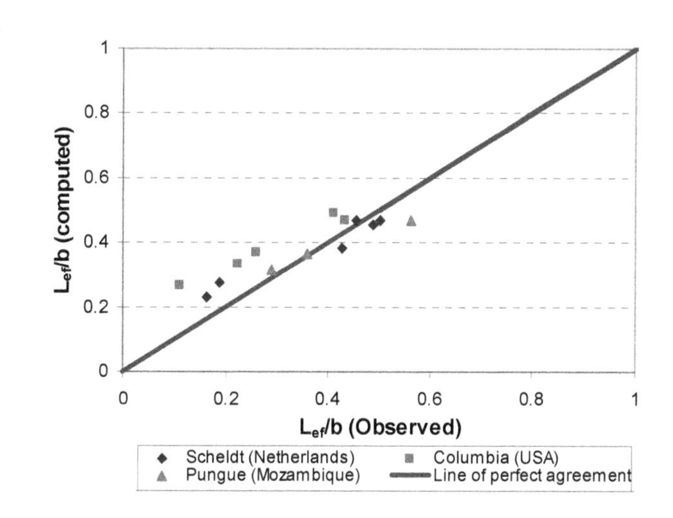

Figure 6.5 Parity diagram for loop lengths in ebb-flood systems of three estuaries: Western Scheldt, Columbia and Pungue.

We can see in Fig. 6.5 that Eq. 6.8 provides a reasonable fit to the observations in the three estuaries, especially for the Western Scheldt and the Pungue. This equation provides a predictive value for the loop length, which is an important parameter for determining the dispersion coefficient. Eq. 6.8 has been derived under the assumption that the loop length obeys a similar equation as the effective longitudinal dispersion presented in Eq. 3.4, Chapter 3.

Conceptual model and analytical equation for the dispersion coefficient

Earlier we have introduced the simplified numerical model to analyze effects of residual circulation on salinity distribution in a hypothetical estuary with a distinct ebb-flood channel system and to investigate the relation between the longitudinal salt transport and relevant parameters. It has been concluded that there is a direct relation between the longitudinal salt transport and the ebb-flood loop length, as well as the tidal pumping efficiency. Therefore, the loop length and the tidal pumping efficiency are likely the two most important parameters for the longitudinal salt transport.

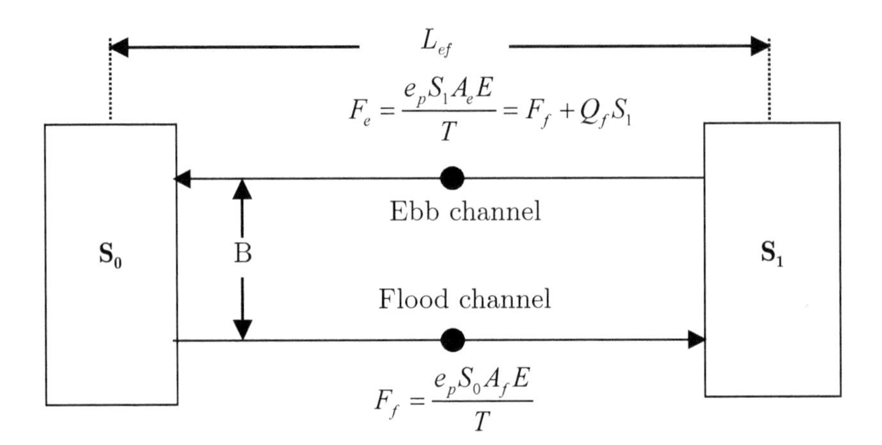

Figure 6.6 Conceptual model for ebb-flood channel dispersion

Based on these findings, a conceptual model has been made to compute the longitudinal dispersion caused by the large-scale residual circulation in an ebb-flood channel estuary. Considering one single loop and looking at the flood channel cross-section over one entire tidal cycle (see Fig. 6.6), the salt flux F_f (L^3T^{-1}) that is conveyed to an upstream cell equals:

$$F_f = S_0 A_f \frac{E}{T} e_p \tag{6.9}$$

where S_0 (-) is the salinity at the beginning of the loop, A_f (L^2) is the cross-sectional area of the flood channel, E (L) is the tidal excursion, and T (T) is the tidal period. e_p (-) is the tidal pumping efficiency defined by Eq. 6.7.

Assuming that the pumping efficiency in the flood and ebb channel is the same, we can determine the salt flux F_e (L^3T^{-1}) in downstream direction through the ebb channel as:

$$F_e = -S_1 A_e \Delta v_e = -S_1 A_e \frac{E}{T} e_p \tag{6.10}$$

in which A_e (L^2) is the cross-sectional area of the ebb channel and S_1 (-) is the salinity at the end of the loop.

In order to close the salt balance, under a steady-state situation (Ippen and Harleman, 1961), the sum of salt fluxes F_f and F_e should equal the salt flux caused by freshwater discharge.

$$-S_1 A_e \frac{E}{T} e_p + S_0 A_f \frac{E}{T} e_p = -Q_f S_1 \tag{6.11}$$

Assuming that the cross-sectional areas in the ebb and flood channel are equal, then Eq. 6.11 becomes:

$$A_f \frac{E}{T} e_p \left(S_1 - S_0\right) = Q_f S_1$$

or:

$$A_f \frac{E}{T} e_p L_{ef} \frac{\Delta S}{\Delta x} = Q_f S \tag{6.12}$$

If the salinity gradient $\mathrm{d}S / \mathrm{d}x$ may be considered constant over the loop, Eq. 6.12 corresponds with the steady state salt transport equation (see Ippen and Harleman, 1961 or Savenije, 1989):

$$AD \frac{\mathrm{d}S}{\mathrm{d}x} = Q_f S \tag{6.13}$$

Combining Eq. 6.12 and Eq. 6.13, we obtain the expression for the effective longitudinal dispersion resulting from ebb-flood channel interaction:

$$D_{ef} = \frac{A_f}{A} \frac{E}{T} e_p L_{ef} \tag{6.14}$$

in which D_{ef} (L^2T^{-1}) is the effective longitudinal tidal average dispersion coefficient caused by the ebb-flood channel interaction.

Equation 6.14 is analogous to the equation introduced by Arons and Stommel (1951) and Zimmerman (1976, 1981), (i.e. $D = k_A U_0 l_0$), where $k_A \simeq e_p \left(A_f / A\right)$, $U_0 \simeq E / T$ and $l_0 \simeq L_{ef}$. In the newly developed equation, the tidal-driven mixing is determined by the pumping efficiency, the tidal excursion and the loop length. The shape of the estuarine geometry (i.e. the loop length and width) plays a significant role in determining the ebb-flood channel residual transport dispersion and this agrees well with the findings of Zimmerman (1981), who concluded that the shape of the coastal geometry or the bottom morphology is important to any tidal residual flow mechanism.

Combination of Eq. 6.14 and Eq. 6.8 yields:

$$D_{ef} = \frac{A_f}{A} \frac{E}{T} e_p \alpha_L \left(1 - \frac{B_L}{B_0 - B_L} \left(\frac{B_0}{B} - 1\right)\right) b \tag{6.15}$$

Assuming an exponential shape of the estuary (see Chapter 3), we obtain:

$$D_{ef} = \varphi_p \frac{Eb}{T} \left(\frac{1 - \dfrac{B_L}{B_0} \exp\left(\dfrac{x}{b}\right)}{1 - \dfrac{B_L}{B_0}} \right) \tag{6.16}$$

in which $\varphi_p = \alpha_L e_p \left(A_f / A\right)$ is the scaling coefficient, which depends on the pumping efficiency of the estuary.

6.3 Results

6.3.1 Residual circulation characteristics of the Western Scheldt obtained from the "virtual laboratory" and the conceptual model

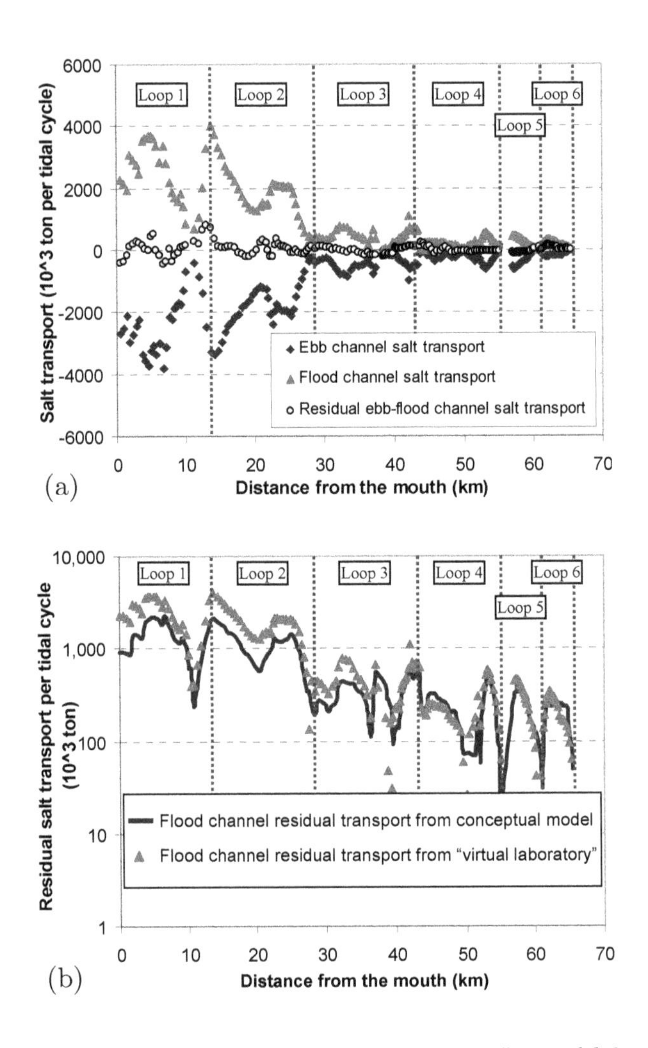

Figure 6.7 (a) Computation of residual salt transport using the "virtual laboratory" data; (b) Comparison between the conceptual box model and the "virtual laboratory" in determining the flood channel residual transport.

Using the "virtual laboratory" output and applying the method of Uncles *et al.* (1985), we are able to compute the residual salt transport for the entire estuary. The computed results are presented in Fig. 6.7a. One can see that over one tidal cycle, there is indeed an upstream residual salt transport in the flood channels and a downstream transport in the ebb channels. This confirms the assumption underlying in the conceptual model.

If the estuary is in a steady state, the net salt transport over the cross-section equals zero. As we can see in Fig. 6.7a, the mean value of the net salt transport over the cross section is close to zero. Near the estuary mouth, there is a large deviation. This may be due to the sea boundary condition, irregular topographies or non-perpendicular flows close to the estuary mouth. We can see in Fig. 6.1 that especially in the first two loops, there are connecting channels and higher order meanders, which may cause this strong fluctuations. Given the large momentary fluxes in the ebb and flood branches, the residual flux is expected to have a high relative error.

A comparison between the conceptual model and the "virtual laboratory" results for F_f (i.e. residual salt transport over one tidal cycle in flood channels) is presented in Fig. 6.7b. The main input data used were: Area of the flood channel A_f (from topographical data); S_0 (salt content at upstream end of the loop) and e_p (pumping efficiency) estimated with Eq. 6.7 using the velocities computed by the hydrodynamic model. From Fig. 6.7b, we can conclude that the conceptual model results computed with Eq. 6.9 and the "virtual laboratory" results fit reasonably well. One can see that jumps occur at the crossover points of the loops and these are generally spread over a certain distance with connecting channels.

6.3.2 Comparison between the conceptual model, the salt intrusion model and the virtual laboratory in computing the dispersion coefficient

We use three methods to compute the dispersion value: (i) the conceptual model and its analytical equation (Eq. 6.16); (ii) the "virtual laboratory" based on the 3-D hydrodynamic model; and (iii) the salt intrusion model of Savenije (1989, 1993c). For details on the salt intrusion model, reference is made to Savenije (1993c) as well as Chapter 3 of this thesis. A comparison among these models when applied to determine the tidally averaged longitudinal dispersion coefficient is presented in Fig. 6.8. The tidal pumping efficiency coefficient and other relevant parameters used in Eq. 6.16 have been obtained from the simulation of the hydrodynamic model as well as from observations.

The dispersion coefficient values computed by the salt intrusion model reflect the "total" effective dispersion, i.e. the sum of horizontal circulation and gravitational circulation (curve number 1 in Figs. 6.8a and 6.8b). The values computed by the conceptual box model only refer to the dispersion by the horizontal ebb-flood channel circulation D_{ef} (i.e. D(Pump), curve number 2 in Figs. 6.8a and 6.8b). The "virtual laboratory" dispersions for tidal pumping and gravitational circulation are shown in curves number 4 and 5 in the same figures, respectively.

Between 10 and 25 km from the mouth, the D_{ef} curve does not agree the values computed by the tidal pumping virtual laboratory (curve number 4). However, the agreement is good in the other reaches. After subtracting the D_{ef} curve from the curve of the total dispersion (i.e. curve number 1), we obtain a curve for gravitational circulation, which is presented by curve number 3. One interesting thing is that, although there is considerable scatter, curve number 3 reasonably fits the gravitational dispersion values computed by the virtual laboratory.

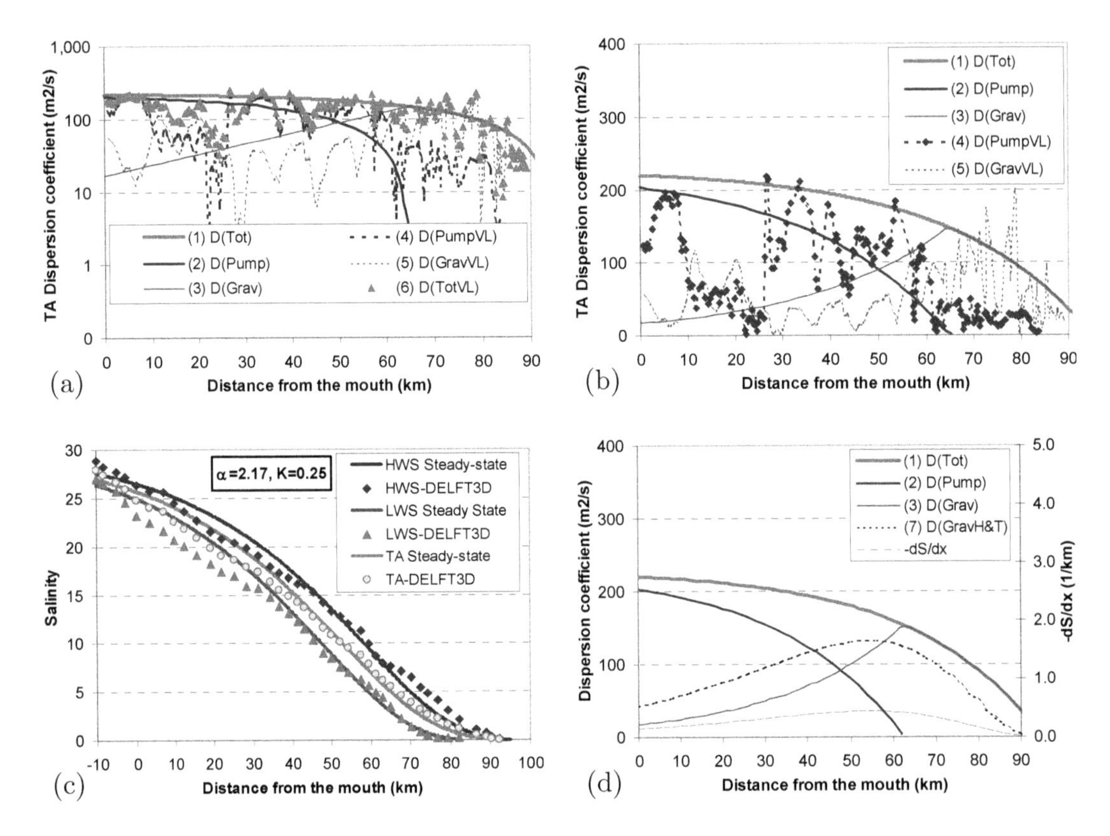

Figure 6.8 Salinity and dispersion in the Western Scheldt: (a) log-scale tidal average dispersion a function of x: Comparison between conceptual model, salt intrusion model and "virtual laboratory" results; (b) same as (a) on normal-scale; (c) Salinity distribution from "virtual laboratory" and salt intrusion model; (d) same as (b) compared to gravitational dispersion computed by Thatcher and Harleman and -dS/dx. Legends: (1) D(Tot): Dispersion computed by salt intrusion model; (2) D(Pump): Dispersion caused by ebb-flood channel circulation (Eq. 6.16); (3) D(Grav): Dispersion caused by gravitational circulation, D(Grav) = D(Tot) - D(Pump); (4) D(PumpVL): Virtual laboratory tidal pumping dispersion; (5) D(GravVL) Virtual laboratory gravitational circulation dispersion; (6) D(TotVL): Virtual laboratory total dispersion, D(TotVL) = D(PumpVL) + D(GravVL); and (7) D(GravH&T): Halerman-Thatcher's gravitation dispersion.

Looking at Figs. 6.8a and 6.8b, the dispersion values computed by the virtual laboratory indicate that at the downstream (seaward) reach of the estuary, the gravitational circulation plays a minor role in producing dispersion. As expected, these figures confirm that dispersion by the residual ebb-flood channel circulation is dominant in the seaward part of a well-mixed ebb-flood system estuary. However, starting from 60 km from the mouth (middle of the estuary), it appears that not only the residual circulation but also gravitational circulation becomes an important mixing mechanism. Upstream from the 65 km mark, where the separation of ebb and flood channels no longer exists, dispersion by residual ebb-flood channel circulation disappears and dispersion by gravitational circulation is the major mechanism. The relative importance of gravitational circulation can hardly be seen on the log scale of Fig. 6.8a, but it can be clearly seen at the linear scale of Fig. 6.8b.

Although scattered and fluctuating over the considered reach, the total dispersion values (red triangles in Figs. 6.8a) computed by the virtual laboratory agree with the results of the salt intrusion model. This agreement allows us to conclude that despite the different approaches applied to the system, the overall result in determination of the dispersion is similar.

It is remarked that when applying the method of Uncles *et al.* (1985) to compute dispersion values, the "virtual laboratory" data obtained from the hydrodynamic model give some unexpected small values for the tidal pumping dispersion between 10 and 25 km from the mouth and unexpected high values for the gravitational dispersion at this location (see Figs. 6.8a and 6.8b). The relatively high gravitational dispersion and low tidal pumping dispersion may be due to the difficulty of the "virtual laboratory" to reproduce the salinity field (see Fig. 6.8c) or due to the limitations of the decomposition method (see Section 2.3.2, Chapter 2). It may also due to the interaction between flood and ebb channels having different salinity or due to effects of the SIPS (i.e. Strain-Induced Periodic Stratification) mechanism (Simpson *et al.*, 1990), tidal trapping (Jay and Musiak, 1994) and effects of junction (Abraham *et al.*, 1986).

The conceptual model does not provide a solution to determine the gravitational circulation. In order to obtain a complete solution, we decide to use the method of Harleman and Thatcher (1974) to compute this type of dispersion. Harleman and Thatcher introduced a relationship for density driven (gravitational) and geometric dispersion:

$$D = D_{\Delta\rho} + D_g \tag{6.17}$$

$$D_{\Delta\rho} = m_1 E_D^{-\frac{1}{4}} \upsilon L \frac{\partial \frac{\overline{\rho}}{\Delta\rho}}{\partial \frac{x}{L}} \tag{6.18}$$

$$D_g = m_2 20 R u_* \tag{6.19}$$

in which $D_{\Delta\rho}$ (L^2T^{-1}) is the coefficient for density driven dispersion and D_g (L^2T^{-1}) is the coefficient for geometry driven dispersion. In these equations: m_1 and m_2 are dimensionless coefficients, υ (LT^{-1}) is the tidal velocity amplitude, u_* (LT^{-1}) is the shear velocity, L (L) is the length of the salt intrusion, R (L) is the hydraulic radius and E_D (-) is the estuary number defined in Chapter 2, Section 2.2.4.

The results are plotted in Figs. 6.8d. We can see that for the cases of the Western Scheldt, the agreement is not very good. Particularly, the maximum value of gravitational circulation dispersion computed by the method of Harleman and Thatcher appears at the 55-km mark, while the maximum value for this dispersion computed by the combination between the conceptual model and the salt intrusion model lies in the 65-km mark, which is the point where the separation of ebb and flood channels no longer exists.

6.3.3 Application to observations in the Western Scheldt and the Pungue

We apply the newly developed equation (i.e. Eq. 6.16) to observations in the Western Scheldt and the Pungue to validate the expressions for the TA dispersion due to residual ebb-flood channel circulation and gravitational circulation under different discharge regimes.

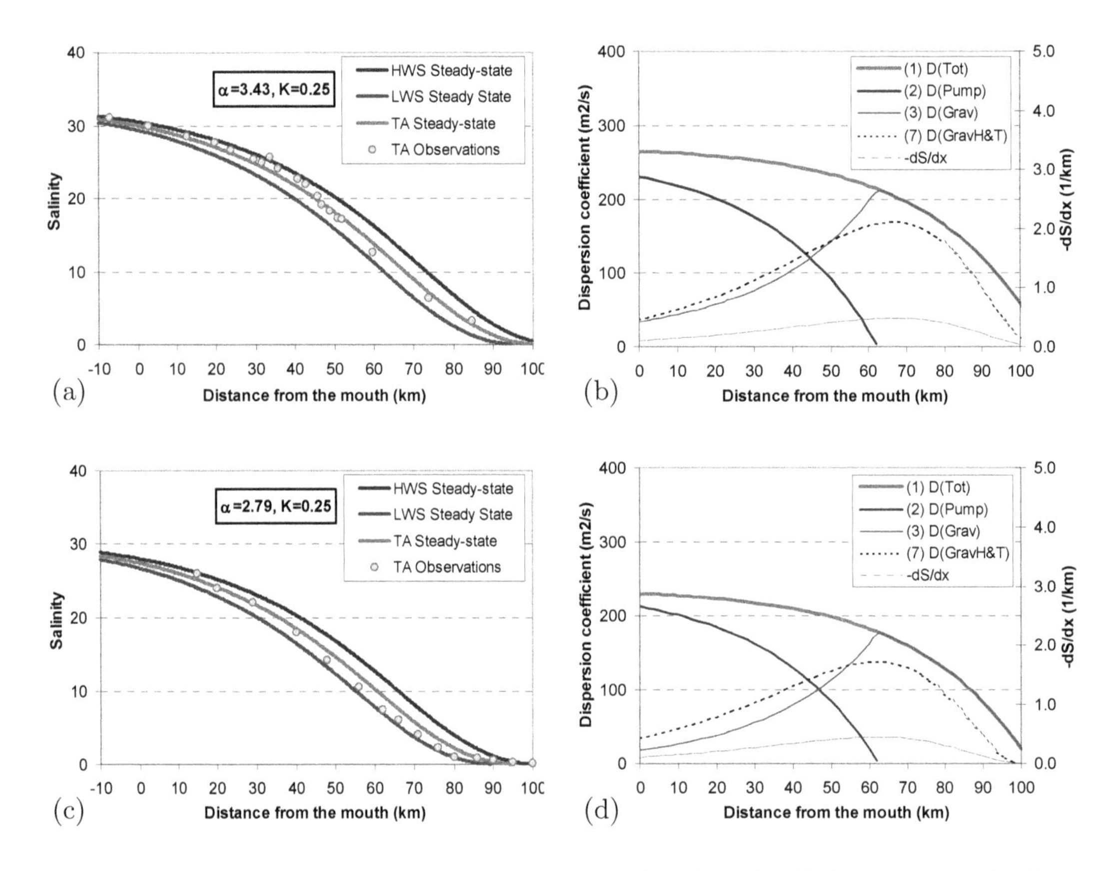

Figure 6.9 Salinity computation in the Western Scheldt: (a) Modelled and observed salinity distribution corresponding with Q_f = 90 m^3/s and (b) TA dispersion coefficient and its components related to (a); (c) Modelled and observed salinity distribution corresponding with Q_f = 95 m^3/s and (d) TA dispersion coefficient and its components related to (c). The legends have already been indicated in Fig. 6.8.

Data of the Western Scheldt have been obtained from Savenije (2005, p. 147), based on observations with Q_f = 90 m^3/s, and from Regnier *et al.* (1998) for the case Q_f = 95 m^3/s. The salt intrusion curves and the longitudinal distribution of the total dispersion, the dispersion by residual circulation and the dispersion by gravitational circulation are presented in Fig. 6.9. In Figs. 6.9b and 6.9d, the dashed line indicates the salinity gradient, which, as expected, has a similar shape as the dispersion by gravitational circulation. The "virtual laboratory" case of the Western Scheldt is also presented in Figs. 6.8c and 6.8d for reasons of comparison. In Figs. 6.9b and 6.9d, the gravitational circulation dispersion computed by the method of Harleman and

Thatcher appears to be in the same shape with the dispersion computed by the combination between the conceptual model and the salt intrusion model, except it underestimates the maximum value.

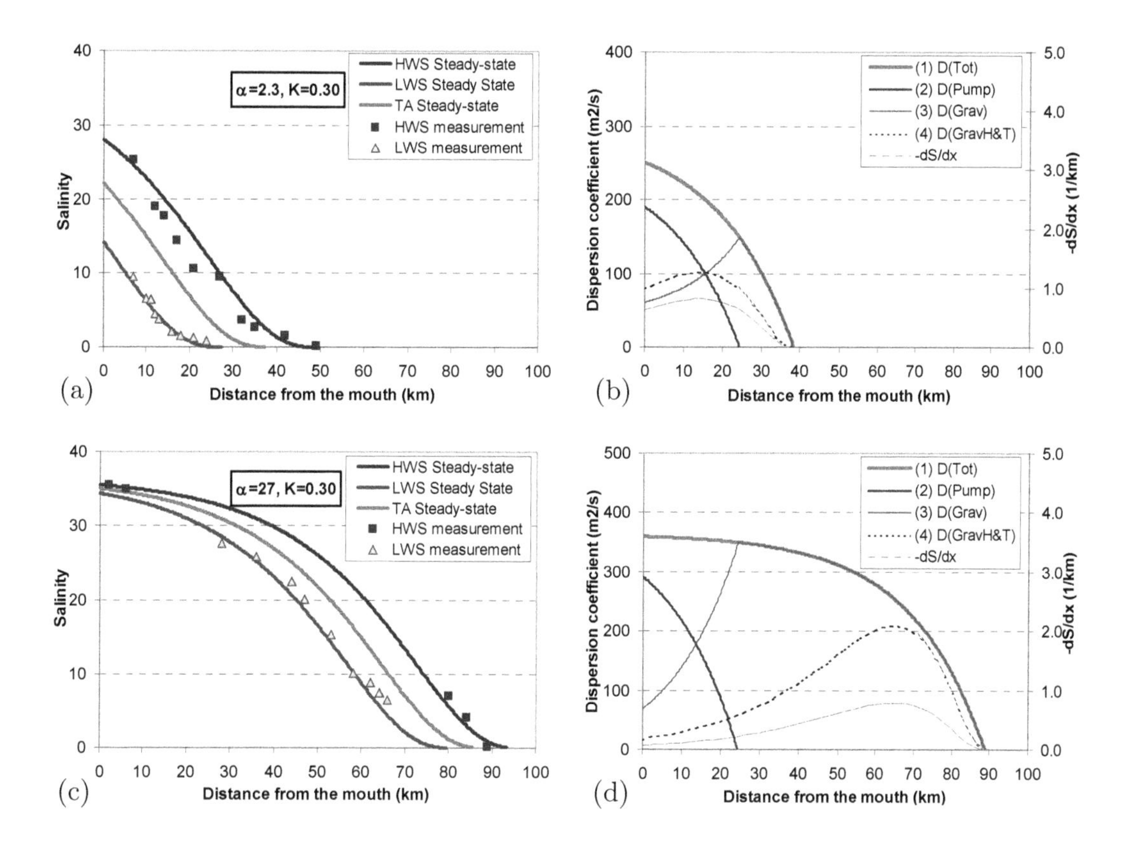

Figure 6.10 Salinity computation in the Pungue: (a) Modelled and observed salinity distribution corresponding with Q_f = 200 m^3/s and (b) TA dispersion coefficient and its components related to (a); (c) Modelled and observed salinity distribution corresponding with Q_f = 20 m^3/s and (d) TA dispersion coefficient and its components related to (c). The legends have already been indicated in Fig. 6.8.

Data of the Pungue estuary have been obtained from Grass (2002), based on observations in 1993 (with Q_f = 20 m^3/s) and 2002 (with Q_f = 200 m^3/s). The salt intrusion curves and the longitudinal distribution of the total dispersion, the dispersion by residual circulation and the dispersion by gravitational circulation are presented in Fig. 6.10. It appear that the computation for the Pungue does not give the satisfactory results in the case of 2002 (Fig. 6.10b), and in general the results of Harleman and Thatcher's method do not agree well with the dispersion computed by the combination between the salt intrusion model and the conceptual model.

The tidal pumping efficiency values obtained are presented in Table 6.2. One can see in this table that for the higher discharge, the tidal pumping efficiency is smaller and the TA salinity intrusion length is smaller as well. It can be explained as follows: Looking at Eq. 6.7, when the river discharge increases, the mean flood velocity in the flood channel decreases and the mean ebb velocity in the flood channel increases. As a result, v decreases, while $\tilde{v}_f$ remains more or less the same. Therefore, the tidal pumping efficiency decreases. A smaller tidal pumping efficiency causes a smaller D_0 (i.e. the dispersion coefficient at the mouth). Combined with a higher discharge, this gives a smaller α_0 (i.e. the mixing coefficient at the mouth), and subsequently, a shorter salinity intrusion length L_{TA} according to the salt intrusion model (see Chapter 3, Section 3.2). This fact is well demonstrated in Table 6.2.

Table 6.2 Tidal pumping efficiency values for the Western Scheldt and the Pungue

	Western Scheldt			Pungue	
	Year 1987 (observations) $Q_f = 90$ m^3/s	Regnier *et al.* (observations) $Q_f = 95$ m^3/s	DELFT3D (virtual laboratory) $Q_f = 120$ m^3/s	Year 1993 (observations) $Q_f = 20$ m^3/s	Year 2002 (observations) $Q_f = 200$ m^3/s
$\frac{\Delta v}{v}$ (-)	12.5×10^{-2}	11.5×10^{-2}	10.5×10^{-2}	19×10^{-2}	9×10^{-2}
φ_p (-)	3.1×10^{-2}	2.9×10^{-2}	2.6×10^{-2}	4.8×10^{-2}	2.3×10^{-2}
L_{TA} (km)	105	100	92	85	34

6.4 DISCUSSION

It appears that the newly developed equation for computing the ebb-flood channel residual circulation dispersion (i.e. Eq. 6.16) performs well, especially at the downstream section of the estuary. Eq. 6.16 is dependent on x (i.e. dispersion is a function of x). This agrees with findings in Ippen and Harleman (1961), Stigter and Siemons (1967), Van der Burgh (1972) and Savenije (2005). However, at the limit of the zone where ebb-flood separation no longer exists, near the point where B=B_L, the equation does not provide a good solution (See Figs. 6.8d, 6.9b and 6.9d). It is understood that after the ebb-flood separation zone, tidal pumping only plays a minor part in producing dispersion and the "virtual laboratory" data confirm this. In this zone, where Eq. 6.16 demonstrates a sudden transition, there should be a smooth transition between tidal pumping dispersion and gravitational dispersion. Further research needs to be done to improve this relationship.

There may be other deficiencies in the approach. Firstly, the model assumes a constant salt gradient within one loop, while in reality the salt gradient can vary. In Fig. 6.8d, it appears that the -dS/dx curve gradually increases upstream until reaching a maximum at 55 km from the estuary mouth, after which it gradually decreases upstream. For the longest loop of the system (i.e. the 2nd loop with a length of 14.6 km), the "virtual laboratory" shows that the -dS/dx value ranges from

1.95×10^{-4} to 2.3×10^{-4} between the most upstream and the downstream point of the loop. Secondly, there is a limitation on the length of the tidal excursion in relation to the loop length. For the first three loops of the Western Scheldt and the first loop of the Pungue, the loop length is larger than the tidal excursion, therefore it is correct to assume that water particles only travel forward or backward within the designated channels (i.e. ebb channel and flood channel). However, if the loop length is shorter than the tidal excursion, then it is not correct to assume that all water particles move within the designated channel. Therefore in this case the conceptual model is less accurate. The "tidal random walk" theory of Zimmerman (1976) could possibly be a better interpretation for the mixing mechanism in these short loops.

Throughout this chapter, it has been assumed that only physical processes are responsible for the modeled salinity field in the hydrodynamic model, which in turn is used to estimate the salt dispersion. This poses some limitations for the study. Firstly, the numerical scheme used in the hydrodynamic model has numerical dispersion. However, the numerical dispersion is expected to be minimized due to the use of an Alternating Direction Implicit (ADI) scheme denoted as a "cyclic method" (Stelling and Leendertse, 1991), which is demonstrated to be computationally efficient in Lesser *et al.* (2004). The numerical dispersion therefore is not significant in the area of interests (i.e. the estuarine sections with ebb-flood channel system) due to the large effective dispersion. Secondly, as we mentioned in Section 6.2.1, the computed water levels, discharges and salinities in the model are in good agreement with observations, while velocity measurements taken in the center of a channel are reasonably well reproduced by the model. However, the computed velocity results are not in good agreement with observations on side slopes of channels, where often a transition occurs from ebb-directed to flood-directed residual transport. This is partly caused by the resolution of the model grid, which is not sufficiently high to resolve relatively steep gradients of the bathymetry. In addition, Winterwerp *et al.* (2006) using a grid cell size of approximately 25 m in the area of interest (i.e. at the river bend near Antwerpen, Belgium) indicated that the velocity profiles are very sensitive to the salinity structure. Therefore as we use a coarser grid size, it may be that the structure of the flow field is not fully correct and subsequently, there may be an unsatisfactory representation of the salinity field.

Moreover, the hydrodynamic model does not explicitly include other processes producing residual currents, such as "Tidal random walk" of Zimmerman (1976), or SIPS, which is a mechanism resulting in particle trapping because of the occurrence of tidal velocity asymmetry and its interaction with the time–varying concentration field (Simpson *et al.*, 1990), or nonlinear effects (Tee, 1978). However, the model computes the entire velocity field and in principles implicitly takes account of all these processes. Whether this is fully in agreement with reality may be questioned but the model has been calibrated to correctly reproduce the velocity field, including residual circulation, adequate enough for morphological computation, for which purpose the model was originally made. In the model, we also do not include the effect of wind, therefore freshwater may remain too long near the downstream boundary without being replaced by ocean water, resulting in a lower salinity near the sea boundary (see Fig. 6.8c).

The performance of the new conceptual model when applied to the "virtual laboratory" and observations in the Western Scheldt provides a promising picture. We consider that the conceptual model's equation is indeed applicable in the case of the Western Scheldt, but it may not yet be a good solution for general applications as we can see in Fig. 6.10 for the case of the Pungue. The equation is definitely a function of the identified parameters, such as the tidal pumping efficiency, the tidal excursion and the estuary shape, particularly the width. However, the right form of the equation needs to be investigated further when we have proper data sets from real world observations.

The newly developed equation does not account for gravitational circulation. The method of Harleman and Thatcher provides a reasonable estimate for this type of dispersion. However, it is noted that this method takes into account mostly vertically gravitational dispersion and this is the reason why we do not obtain a very good correspondence with the subtraction of the total dispersion and the ebb-flood channel dispersion (Figs. 6.8d, 6.9b, 6.9d, 6.10b and 6.10d). Nevertheless, it is advised to consider the newly developed equation together with the method of Harleman and Thatcher in order to obtain the full dispersion in an ebb-flood channel estuary.

The dispersion values computed from the virtual laboratory appear to agree fairly well with the other two models (i.e. the conceptual model and the salt intrusion model) for the case of the Western Scheldt. When computing the dispersion coefficient on the basis of the observed salinity distribution, the values obtained depend on the salinity gradient. This gradient is very sensitive to errors in the observed salinities at subsequent grid points. As a result, we notice that a small variation in the observed salinity values along the estuary can lead to large fluctuations in the computed dispersion values, as one can see in Fig. 6.8a and 6.8b. The average values, however, follow the main pattern. Furthermore, it is possible that the relative error that we make when subtracting fluxes can be very large, particularly if the residual fluxes are small compared to the momentary fluxes.

6.5 CONCLUSIONS

This chapter presents a study on the tidal pumping effect caused by the large-scale residual circulation through ebb and flood channels. It has been demonstrated that in the seaward part of converging estuaries such as the Western Scheldt and the Pungue, where the density gradient is small, this kind of tidal pumping is the main mixing mechanism.

Based on a conceptual model, a new equation determining the tidal pumping dispersion has been derived. The new equation provides an opportunity to evaluate the large-scale mixing mechanism caused by the residual ebb-flood channel circulation in terms of a dispersion coefficient, which is not feasible to obtain from field observations. The equation takes into account two important parameters of the residual circulation, i.e. the tidal pumping efficiency and the ebb-flood loop length. The new equation has subsequently been compared with "virtual laboratory" data of the Western Scheldt estuary in the Netherlands generated by the DELFT3D hydrodynamic model and with the steady-state salt intrusion model's results. The comparison confirms an agreement between the new equation and the existing models in determining the residual transport and the tidal pumping dispersion coefficient. Finally, the equation has been applied to observations in the Western Scheldt and the Pungue estuary. The good performance of the newly developed equation in comparison with observations in the Western Scheldt indicates that it is applicable in practice. However, the performance of the equation compared to observations in the Pungue appears less convincing, suggesting that it may not yet be a good solution for the general application, therefore further research needs to be carried out to improve the equation. It is recommended to further study and apply this equation to other ebb-flood channel estuaries, such as the Columbia, several estuaries in the Chesapeake Bay in U.S., and the Thames in U.K., in order to get new insights.

Chapter 7

CONCLUSIONS AND RECOMMENDATIONS

The aim of this study was to address a number of knowledge gaps related to salt intrusion, mixing and tides in multi-channel estuaries. The objectives of this research were formulated as: (i) to investigate and to develop a predictive steady state salt intrusion model for a multi-channel estuary; (ii) to develop a new method for estimating the distribution of freshwater discharge over the branches of a multi-channel estuary; (iii) to analyse characteristics of tidal waves in multi channel estuaries; and (iv) to develop a theory analysing effects of tidal pumping caused by residual ebb-flood channel circulation on salinity distribution in multi-channel estuaries and to propose a new analytical equation quantifying the 1-D effective salt dispersion coefficient for the tidal pumping mechanism. With the results presented in the previous chapters, these objectives have been met. As a result, a number of conclusions and recommendations have been formulated, which are presented below.

7.1 CONCLUSIONS

Investigation and development of a predictive steady state salt intrusion model for a multi-channel estuary

For a 1-D predictive model, one needs adequate hydraulic parameters and a predictive theory on mixing processes as a function of the river discharge and tidal regime. In particular, the dispersion coefficient should be predictable in time and space. To date, a model has not yet been developed to cope with salinity intrusion in multi-channel estuaries such as the Mekong Delta, which consists of eight branches. On the basis of salinity measurements during the dry season of 2005 and 2006, an analytical model, and subsequently, a predictive steady-state model for salt intrusion, based on the theory for the computation of salt intrusion in single alluvial estuaries, have been developed to compute the longitudinal salinity distribution (at HWS and LWS) for the estuary branches of the Mekong Delta. In view of the similar hydraulic, topographical and salinity characteristics of the branches, it is concluded that the multi-channel estuarine system functions as an entity and that paired branches can be considered as a single estuary branch. This procedure has been successfully applied and tested in the Dinh An and Tran De branches, the Co Chien and Cung Hau branches and the Tieu and Dai branches. The model has been validated with data of the dry seasons in 1998 and 1999 for the combined Hau and Co Chien - Cung Hau estuaries. The overall results of the salinity computation indicate that combining branches is both practical and physically sound, while the simplified method produces good results for a complex multi-channel estuary such as the Mekong Delta. In addition, an innovation introduced is that the predictive model can now be used in estuaries where the tidal excursion is not constant, but damped (Mekong) or amplified (Scheldt).

Development of a new method for estimating the distribution of freshwater discharge over the branches of a multi-channel estuary

The determination of the fresh water discharge in estuaries is complicated, as it requires detailed measurements during a full tidal cycle. Moreover, in the dry season when the salt intrusion matters most, the magnitude of the freshwater discharge is small compared to the tidal flow (often within the measurement error of the tidal flow). It is even more difficult to determine the discharge in a complex system such as the Mekong Delta, which consists of eight branches over which the fresh water discharge is distributed. Especially, the distribution of the discharge over the branches in the saline reaches depends on a complex interaction of topography, tide, network layout (hydraulic structures, canals, etc.) and additional withdrawals or drainage. We have seen that the analytical model for salt intrusion in the Mekong Delta can be used to predict the salinity distribution in the estuary branches, if topography, tide and river discharge are known. But the reverse also applies. If the salinity distribution in the Mekong branches is known, we can estimate the fresh water discharge. The discharge of a branch can be computed by the predictive equation (i.e. Eq. 4.2), making use of the two parameters, α_0^{HWS} and E_0, obtained from the predictive salt intrusion model. It appears that the results of the new analytical approach agree well with observations and with the results of a more complex hydrodynamic model and with observations upstream of the system. The study shows that with relatively simple salinity measurements and making use of the predictive salt intrusion model, the new method can provide a correct picture of the discharge distribution over a multi-channel estuary.

Tidal wave characteristics in multi-channel estuaries

Observations in alluvial estuaries indicate that a damped tidal wave moves slower than is indicated by the classical equation for wave celerity, whereas the celerity is higher if the tidal wave is amplified. The tidal damping pattern in an estuary appears in one of three types: amplified, un-damped (ideal) or damped. The phase lag between HW and HWS (and LW-LWS) lies between zero and $\pi/2$. In this study, the tidal wave characteristics of the two multi-channel estuaries, i.e. the Mekong Delta and the Scheldt, have been investigated on the basis of analytical equations, hydrodynamic models and observations.

The Mekong Delta estuaries are mostly riverine in character. They have a small estuary shape number, therefore have a large phase lag and host a damped tidal wave, which moves slower than indicated by the classical celerity equation. It is concluded that in a complex system containing a number of branches, there is a dominant branch, which represents the characteristics of the system as a whole and it forces other branches to adapt and to adjust. For the Tien river system, which is a complex system containing three big branches and five sub-branches, the My Tho river is the dominant branch while being the longest branch. For a less complex system (i.e. the Hau river, which contains only two paired branches of the same length), the analytical equations agree very well with observations and with the calibrated hydrodynamic model. The agreement between the analytical equations, hydrodynamic model and observations in the Tien river system appears less convincing. Therefore, it is concluded that the analytical equations do not well represent for a complex multi-channel system such as the Tien river; while the hydrodynamic model may require more detailed data on topography and hydrology. The junctions between branches play a significant role in changing the pattern of tidal range, phase lag and tidal wave propagation. The calibration of the hydrodynamic model should be done with special care to the junction of branches.

The Scheldt estuary is a more marine estuary with a large estuary shape number, a small phase lag and an amplified wave. The tidal wave propagation in the Scheldt estuary is faster than suggested by the classical celerity equation due to the amplification of the tidal range. It is concluded that the tidal wave celerity in the Scheldt estuary depends mainly on the topography. The analytical equations are very well able to describe the tidal dynamics in the Scheldt, which has a complex structure of ebb and flood channels.

Development of a new theory for analysing effects of tidal pumping caused by residual ebb-flood channel circulation in multi-channel estuaries and a new analytical equation quantifying the 1-D effective salt dispersion coefficient due to tidal pumping

In the seaward part of converging estuaries with a distinct ebb-flood channel system, such as the Western Scheldt and the Pungue, where the density gradient is small, tidal pumping caused by the residual circulation through ebb and flood channels is the main mixing mechanism. Results obtained from the schematisation of a hypothetical ebb-flood channel estuary reveal that there is a direct relation between the longitudinal salt transport and the ebb-flood loop length, as well as the tidal pumping efficiency. Therefore, the loop length and the tidal pumping efficiency are likely the two most important parameters for the residual circulation mixing. On the basis of these parameters, a new equation for dispersion by ebb and flood channel interaction has been derived. A 3-D hydrodynamic model in DELFT3D has been employed as "virtual laboratory" to generate a solution of the circulation pattern for the Western Scheldt estuary. This solution has been subsequently decomposed to isolate the influence of different mixing hydrodynamic processes. The new analytical equation has been compared with the decomposition results and reasonable agreement has been found. Mixing caused by gravitational circulation is not estimated by the analytical equation, but this can be addressed by the method of Harleman and Thatcher (1974). Observations from the Western Scheldt estuary and the Pungue have been used to validate calculated dispersion values. The good performance of the newly developed equation in comparison with the existing models as well as with observations indicates that the equation is indeed applicable in practice. The new equation provides an opportunity to evaluate the large-scale mixing mechanism caused by the residual ebb-flood channel circulation, which is not feasible to analyse from field observations.

7.2 RECOMMENDATIONS

Predictive salinity model and discharge predictive equation for multi-channel estuaries

- It is suggested to incorporate the analytically predictive function of the dispersion coefficient into hydrodynamic models. The analytical model can provide the dispersion as a function of time and space. As a result, a new generation of dynamic water quality models will be developed that are predictive without the need for detailed information.
- It is recommended to further apply the predictive salinity model and the freshwater discharge predictive equation to other multi-channel estuaries in the world, e.g. the Loire (France), the Tanintharyi (Myanmar), the Dhamra (India) or the Yangtze (China). These applications will yield useful management information, while at the same time new insights may be obtained.
- It is recommended to carry out complete salinity measurements by moving boat method for seven branches of the Mekong Delta within the same day, preferable during spring tide in the dry season, when the salt intrusion matters most and the system is close to a steady state. These measurements shall provide a good overview of how salinity and discharge distribute over the branches of the Mekong.

Tidal wave characteristics in multi-channel estuaries

- More detailed field observations of the longitudinal tidal damping and phase lag, which are two important factors for classifying estuaries, are recommended to be carried out in the two study areas. The measurements will provide more confidence in the method used in this study and improve the performance of the set of analytical equations and the hydrodynamic models.
- It is recommended that the analytical equations can be used to facilitate the calibration of hydrodynamic models. Especially for rivers and estuaries where

there are not many observations available, the set of the analytical equations is a good tool to interpolate between observations and to check the reliability of a hydrodynamic model. However, further improvement of the phase lag equation (i.e. Eq. 2.51) is needed by taking into account effects of the freshwater discharge and the bottom slope.

- Sudden jump in tidal wave propagation occurring at the junction point (for example with the case of the Tien river system in the Mekong Delta) is a very complicated phenomenon. It could be useful to have observations and to employ a "particle-tracking" model to analyse this phenomenon.

Mixing in estuaries with a distinct ebb-flood channel system

- The use of the 3-D “virtual laboratory” data (i.e. model-derived data) creates a unique chance to analyse large–scale mixing mechanisms such as tidal pumping caused by ebb-flood channel residual circulation. It is recommended to make more use of model-derived data to obtain insights in circulation patterns, whereby it is suggested to concentrate on the larger scale processes, since small scale processes show large fluctuations, mostly due to the inadequacy to correctly represent small-scale topography in the models.
- It is suggested that the newly developed equation (i.e. Eq. 6.16) may not yet be a good solution for the general application, therefore further research needs to be carried out to improve the equation. It is also recommended to further study and apply the equation to other ebb-flood channel estuaries, such as the Columbia, several estuaries in the Chesapeake Bay in U.S., and the Thames in U.K., in order to get new insights.
- At the limit of the zone where ebb-flood separation no longer exists, near the point where $B = B_L$, the new equation does not provide a good solution. It is understood that after the ebb-flood separation zone, tidal pumping only plays a minor part in producing dispersion and the "virtual laboratory" data confirm this. In this zone, where the equation demonstrates a sudden transition, there should be a smooth transition between tidal pumping dispersion and gravitational dispersion. Further research needs to be done to refine this transition.
- The new equation does not account for gravitational circulation and therefore it has to be combined with other approaches in order to obtain a complete solution for the dispersion coefficient. The method of Harleman and Thatcher (1974) provides a good starting point for this type of dispersion. However, it will be necessary to refine the method to better account for the large-scale and small-scale topography, and in particular for estuaries with a distinct ebb-flood channel system.

REFERENCES

Abbott, M.B., and Basco, D.R., 1989. Computational Fluid Dynamics. Longman Scientific & Technical, Longman Group UK Limited, 425 pp.

Abraham, G., De Jong, P. and Van Kruiningen, F.E., 1986. Large-scale mixing processes in a partly mixed estuary. In: J. Van de Kreeke (ed.), Physics of shallow estuaries and bays. Springer, Berlin, pp. 6-21.

Abraham, G., Karelse, M., and Lases, W., 1975. Data requirements for exchange coefficients for non-homogeneous flow. Proceedings of the 16th International Association Hydraulic Research Congress, Vol. 3, pp. 275-283.

Ahnert, F., 1960. Estuarine meanders in the Chesapeake Bay area. Geographical Review 50, 390–401.

Arends, A.A., and Winterwerp, J.C., 2001. Morphological management concept for the Western Scheldt - part of the long-term vision on the Scheldt-estuary. Proceedings of the Conference on Coastal Zone management, COZU01.

Arons, A. B. and Stommel, H., 1951. A mixing-length theory of tidal flushing. Transactions, American Geophysical Union 32, 419-421.

Aubrey, D. G., and Speer, P. E., 1985.A study of non-linear tidal propagation in shallow inlet/estuarine systems, Part I: Observations. Estuarine, Coastal and Shelf Science 21, 185-205.

Baeyens, W., Van Eck, G., Lambert, C., Wollast, R., and Goeyens, L., 1998. General description of the Scheldt estuary. Hydrobiologia 366, 1-14.

Biggs, R.B., and Cronin, L.E., 1981. Special characteristics of estuaries. In B. J. Neilson and L. E. Cronin (eds.), Estuaries and nutrients , Humana., pp. 3-23.

Blumenthal, K.P., Abraham, G., Langeweg, F., Weerden, J.J., van Vreugdenhill, C.B., Kolkman, P.A., and Schoenveld, J.C., 1976. Salt distribution in estuaries. Proceedings of Technical meeting 30, Committee for Hydrological Research TNO, The Hague, The Netherlands.

Bowden, K.F., 1963. The mixing processes in a tidal estuary. International Journal of Air and Water Pollution 7, 343–356.

Brockway, R., Bowers, D., Hoguane, A., Dove, V., and Vassele, V., 2006. A note on salt intrusion in funnel-shaped estuaries: Application to the Incomati estuary, Mozambique. Estuarine, Coastal and Shelf Science 66, 1-5.

Cameron, W. M., and Pritchard, D. W., 1963. Estuaries. In: M. N. Hill (editor), The Sea vol. 2. John Wiley and Sons, New York, pp. 306 - 324.

Cancino, L., and Neves, R. J. J., 1994. 3D numerical modelling of cohesive suspended sediment in the Western Scheldt estuary (The Netherlands). Netherlands Journal of Aquatic Ecology 28(3-4), 337- 345.

Cartwright, D.E., 1999. Tides: a scientific history. Cambridge University Press, Cambridge, UK, 292 pp.

Cogels, O., 2005. A regional cooperation programme for sustainable water resources development of the Mekong river basin. Proceedings of the International Symposium on Sustainable development in the Mekong river basin, Ho Chi Minh City, Vietnam, pp. 1-7.

Cunge, J.A., Holly, F.M., Jr., and Verwey, A., 1980. Practical aspects of computational river hydraulics. Pitman Publishing Limited, London (reprinted 1986 by the Iowa Institute of Hydraulic Research, Iowa City), 420 pp.

Dalrymple, R.W., Zaitlin, B.A., and Boyd, R., 1992. Estuarine facies models: conceptual basis and stratigraphic implications. Journal of Sedimentary Research 62, 1130–1146.

Davies, L.J., 1964. A morphogenic approach to the worlds' shorelines. Z. Geomorph. 8, 127-142.

Delft Hydraulics, 2006. SOBEK Flow: Technical reference. Delft Hydraulics, Delft, The Netherlands.

Dronkers, J.J., 1964. Tidal computations in rivers and coastal waters. North-Holland Publishing Company, Amsterdam, 518 pp.

Dronkers, J.J., 2005. Dynamics of coastal systems. World Scientific Pub. Co. Inc, 540 pp.

Dyer, K.R., 1973. Estuaries, a physical introduction. John Wiley, London, 140 pp.

Dyer, K.R., 1995. Sediment transport processes in estuaries. In: G.M.E. Perillo (ed.), Geomorphology and sedimentology of estuaries. Elsevier, Amsterdam, pp. 423 – 447.

Dyer, K.R., 1997. Estuaries, a physical introduction, second edition. John Wiley, London, 195 pp.

Fairbridge, R.W., 1980. The estuary: its definition and geodynamic cycle. In: E. Olausson, I. Cato (Eds.), Chemistry and biogeochemistry of estuaries. John Wiley, Chichester, pp. 1–35.

Fischer, H.B., 1972. Mass transport mechanisms in partially stratified estuaries. Journal Fluid Mech. 53, 672-687.

Fischer, H.B., 1974. Discussion of 'Minimum length of salt intrusion in estuaries' by B.P. Rigter, 1973. Journal Hydraul. Div. Proc. 100, 708-712.

Fischer, H.B., List, E.J., Koh, R.C.Y., Imberger, J., and Brooks, N.H., 1979. Mixing in inland and coastal waters. Academic Press, New York, 483 pp.

Friedrichs, C.T., and Aubrey, D.G., 1988. Non-linear tidal distortion in shallow well-mixed estuaries: a synthesis. Estuarine, Coastal and Shelf Science 27, 521-545.

Friedrichs, C.T., and Aubrey, D.G., 1994. Tidal propagation in strongly convergent channels. Journal of Geophysical Research 99(C2), 3321-3336.

Geyer, W.R., and Signell, R.P., 1992. A reassessment of the role of tidal dispersion in estuaries and bays. Estuaries 15(2), 97-108.

Graas, S., 2002. Improved model for the salt intrusion in the Pungue estuary. Project report, Administracao Regional de Aguas do Centro, Ministrerio das Obras Publicas e Habitacao, Mozambique.

Godin, G., 1999. The propagation of tides up rivers with special considerations on the upper Saint Lawrence river. Estuarine, Coastal and Shelf Science 48, 307-324.

Gross, E. S., Koseff, J. R., and Monismith, S. G., 1999. Evaluation of advective schemes for estuarine salinity simulations. Journal of Hydraulic Engineering 125, 32-46.

Haas, J., 2007. Phase lags in alluvial estuaries: classification of alluvial estuaries by means of the phase lags. M.Sc. thesis. Delft University of Technology, The Netherlands, 156 pp.

Harleman, D.R.F., 1966. Tidal dynamics in estuaries part II : real estuaries. In A.T. Ippen (ed.), Estuarine and Coastal Hydrodynamics. McGraw Hill, New York, USA, pp. 522-545.

Harleman, D.R.F., and Abraham, G., 1966. One-dimensional analysis of salinity intrusion in the Rotterdam Waterway. Delft Hydraulics Laboratory Publication 44, Delft, The Netherlands.

Harleman, D.R.F., and Thatcher, M.L., 1974. Longitudinal dispersion and unsteady salinity intrusion in estuaries. La Houille Blanche 1, 25-33.

Hayes, M.O., 1975. Morphology of sand accumulation in estuaries. In: L.E. Cronin (ed.), Estuarine Research, Vol. 2. Academic Press, New York, pp. 3–22.

Horrevoets, A.C., Savenije, H.H.G., Schuurman, J.N., and Graas, S., 2004. The influence of river discharge on tidal damping in alluvial estuaries. Journal of Hydrology 294(4), 213-228.

Imasato, N., 1983. What is tide-induced residual current?. Journal of Physical Oceanography 13, 1307-1317.

Ippen, A.T., and Harleman, D.R.F., 1961. One-dimentional analysis of salinity intrusion in estuaries. Technical Bulletin number 5, Committee on Tidal Hydraulics, U.S. Army Corps of Engineers.

Ippen, A.T., 1966. Estuary and coastline hydrodynamics. McGraw Hill, New York, USA, 744 pp.

Lewis, R.E., 1979. Transverse velocity and salinity variations in the Tees estuary. Estuarine and Coastal Marine Science 8, 317-326.

Jay, D.A., 1991. Green's Law revisited: Tidal long-wave propagation in channels with strong topography. Journal of Geophysical Research 96, 20585-20598.

Jay, D.A., and Musiak, J.D. 1994. Particle trapping in estuarine tidal flows. Journal of Geophysical Research 99, 445–461.

Jay, D.A., Uncles, R.J., Largier, J., Geyer, W.R., Vallino, J., and Boynton, W.R., 1997. A review of recent developments in estuarine scalar flux estimation. Estuaries 20, 262-280.

Jeuken, M.C.J.L., 2000. On the morphologic behaviour of tidal channels in the Westerschelde estuary. Ph.D. thesis. University of Utrecht, The Netherlands, 378 pp.

Ketchum., B.H., 1951. The exchanges of fresh and salt water in tidal estuaries. Journal of Marine Research 10, 18-38.

Kuijper, C., Steijn, R., Roelvink, J.A., van der Kaaij, T., and Olijslagers, P., 2004. Morphological modelling of the Western Scheldt. Report No. Z3648/A1198 prepared for Rijkswaterstaat Rijksinstituut voor Kust en Zee/RIKZ, Delft, The Netherlands.

Lane, A., Prandle, D., Harrison, A. J., Jones, P. D., and Jarvis, C. J., 1997. Measuring fluxes in tidal estuaries: Sensitivity to instrumentation and associated data analyses. Estuarine, Coastal and Shelf Science 45, 433–451.

Lanzoni, S., and G. Seminara, 1998. On tide propagation in convergent estuaries. Journal of Geophysical Research 103(C13), 30793-30812.

Le, S., 2006. Salinity intrusion in the Mekong Delta. Agriculture Publisher, Ho Chi Minh City, Vietnam, 387 pp. (in Vietnamese).

Le, T.V.H., Nguyen, H.N., Wolanski, E., Tran, T.C., and Haruyama, S., 2007. The combined impact on the flooding in Vietnam's Mekong river delta of local man-made structures, sea level rise, and dams upstream in the river catchment. Estuarine, Coastal and Shelf Science 71, 110-116.

Lesser, G.R., Roelvink, J.A., van Kester, J.A.T.M., and Stelling, G.S., 2004. Development and validation of a three-dimensional morphological model. Coastal Engineering 51, 883– 915.

Lewis, R., 1997. Dispersion in estuaries and coastal waters. John Wiley Publisher, Chichester, UK, 312 pp.

Lewis, R.E., 1979. Transverse velocity and salinity variations in the Tees estuary. Estuarine and Coastal Marine Science 8, 317-326.

Lewis, R.E., and Uncles, R.J., 2003. Factors affecting longitudinal dispersion in estuaries of different scale. Ocean Dynamics 53, 197-207.

Lin, B., and Falconer, R. A., 1997. Tidal flow and transport modeling using ULTIMATE QUICKEST scheme. Journal of Hydraulic Engineering 123, 303–314.

McCarthy, R.K., 1993. Residual currents in tidally dominated, well-mixed estuaries. Tellus 45A, 325–340.

Mekong River Commission (MRC), 2004. Study on Hydro-meteorological monitoring for water quality rules in the Mekong river basin: Final report, II (supporting report 2/2), WUP-JICA, Phnom Penh, Cambodia.

Nichols, M.M., and Biggs, R., 1985. Estuaries. In R.A. Davis (ed.), Coastal sedimentary environments. Springer Verlag,New York, pp. 77–186.

Nihoul, J.C.J., and Ronday, F.C., 1975. The influence of the "tidal stress" on the residual circulation. Tellus 27, 484-490.

Nguyen, A.N., and Nguyen, V.L, 1999. Salt water intrusion disaster in Vietnam. Publication of UNDP Project No VIE/97/002, Ho Chi Minh City, Vietnam (in Vietnamese and English), 68 pp.

Nguyen, A.D., and Savenije, H.H.G., 2006. Salt intrusion in multi-channel estuaries: A case study in the Mekong Delta, Vietnam. Hydrology and Earth System Sciences 10, 743-754.

Nguyen, A.D., Savenije, H.H.G., and van der Wegen, M., 2006. Effect of horizontal residual circulation on salinity distribution in estuaries. Geophysical Research Abstracts. European Geosciences Union, Vienna, Austria, Vol. 8, 01259.

Nguyen, A.D., Savenije, H.H.G., Pham, D.N., and Tang, D.T., 2007a. Using salt intrusion measurements to determine the freshwater discharge distribution over the branches of a multi-channel estuary: The Mekong Delta case. Estuarine, Coastal and Shelf Science, doi 10.1016/j.ecss.2007.10.010.

Nguyen, A.D., Savenije, H.H.G., van der Wegen, M., and Roelvink, J.A., 2007b. New analytical equation determining the dispersion coefficient in estuaries with a distinct ebb-flood channel system. Accepted for publication in Estuarine, Coastal and Shelf Science.

Nguyen, A.D., Savenije, H.H.G., Pham, D.N., and Tang, D.T., 2007c. Tidal wave propagation in the branches of a multi-channel estuary: the Mekong Delta case. In V. Penchev, H.J. Verhagen (Eds.), Proceedings of the 4th International Conference PDCE 2007, Varna, Bulgaria, pp. 239-248.

Nguyen, A.D., Savenije, H.H.G., and Wang, Z.B., 2007d. Tidal wave characteristics in a braided ebb-flood channel estuary. In B. Cetin, S. Ulutürk (Eds.), Proceedings of the 8th International Conference MEDCOAST 2007, Alexandria, Egypt, pp. 1283-1294.

Nguyen, A.D., and Savenije, H.H.G., 2007e. New method to determine the freshwater discharge distribution over the branches of a multi-channel estuary. Proceedings of the Fifth International Symposium on Environmental Hydraulics, Tempe, Arizona, USA, 6 pp.

Nguyen, A.D., Savenije, H.H.G., van der Wegen, M., and Roelvink, J.A., 2007f. Mixing in estuaries with a distinct ebb-flood channel. Proceedings of the Fifth International Symposium on Environmental Hydraulics, Tempe, Arizona, USA, 6 pp.

Nguyen, H.N., Tran, T.C., and Ho, N.D., 2000. The applied assistant software HydroGis for modelling flood and mass transport in Low river delta. Proceeding of International European-Asia Workshop 'Ecosystem & Flood 2000', Hanoi, Vietnam.

Park, J.K. and James, A., 1990. Mass flux estimation and mass transport mechanism in estuaries. Limnology and Oceanography, 35(6), 1301-1313.

Parker, B.B., 1991. Tidal Hydrodynamics. John Wiley & Sons, New York, 883 pp.

Pillsbury, G., 1956. Tidal hydraulics. US Corps of Engineers, Vicksburg, USA.

Pino Q, M., Perillo, G.M.E., and Santamarina, P., 1994. Residual fluxes in a cross-section of the Valdivia river estuary, Chile. Estuarine, Coastal and Shelf Science 38, 491–505.

Prandle, D., and Rahman, M., 1980. Tidal response in estuaries. Journal of Physical Oceanography 10, 1552-1573.

Prandle, D., 1981. Salinity intrusion in estuaries. Journal of Physical Oceanography 11, 1311-1323.

Prandle, D., 2003. Relationships between tidal dynamics and bathymetry in strongly convergent estuaries. Journal of Physical Oceanography 33, 2738-2750.

Prandle, D., 2004. Saline intrusion in partially mixed estuaries. Estuarine, Coastal and Shelf Sciences 59(3), 385–397.

Pritchard, D.W., 1954. A study of the salt balance in a coastal plain estuary. Journal of Marine Research 13, 133-144.

Pritchard, D.W., 1955. Estuarine circulation patterns. Proceedings Amer. Soc. Civil Eng. (ASCE) 81, Paper no. 717, 11 pp.

Pritchard, D. W., 1967. What is an estuary, physical viewpoint. In: G. H. Lauf (ed.), Estuaries. American Association for the Advancement of Science, Washington D.C., Publication no. 83.

Regnier P., Mouchet A., Wollast R., and Ronday F., 1998. A discussion of methods for estimating residual fluxes in strong tidal estuaries. Continental Shelf Research 18(13), 1543-1571.

Rigter, B.P., 1973. Minimum length of salt intrusion in estuaries. Journal of the Hydraulic Division, Proceedings of ASCE 99, 1475-1496.

Rodriguez-Iturbe, I., and Rinaldo, A., 2001. Fractual river basins: Chance and self-organisation (first published 1997). Cambridge University Press, New York/Cambridge, 564 pp.

Roelvink, J.A., and van Banning, G.K.F.M., 1994. Design and development of DELFT3D and application to coastal morphodynamics. In: A. Verwey, A.W. Minns, V. Babovic and C. Maksimovic (eds.), Proceedings of Hydroinformatics' 94, volume 1, 451-456.

Roelvink, J.A., and Walstra, D.J.R., 2004. Keeping it simple by using complex models. Proceedings of 6th International Conference on Hydrosciences and Engineering, Brisbane, Australia.

Savenije, H.H.G., 1986. A one-dimensional model for salinity intrusion in alluvial estuaries, Journal of Hydrology 85, 87-109.

Savenije, H.H.G., 1988. Influence of rain and evaporation on salt intrusion in estuaries. Journal of Hydraulic Engineering 114(12), 1509-1524.

Savenije, H.H.G., 1989. Salt intrusion model for high-water slack, low-water slack and mean tide on spreadsheet. Journal of Hydrology 107, 9-18.

Savenije, H.H.G., 1992a. Rapid assessment technique for salt intrusion in alluvial estuaries. Ph.D. thesis, IHE report series, no. 27, International Institute for Infrastructure, Hydraulics and Environment, Delft, The Netherlands.

Savenije, H.H.G., 1992b. Lagrangean solution of St. Venant's equations for an alluvial estuary. Journal of Hydraulic Engineering 118(8), 1153-1163.

Savenije, H.H.G., and Pagès, J., 1992. Hypersalinity, a dramatic change in the hydrology of Sahelian estuaries. Journal of Hydrology 135, 157-174.

Savenije, H.H.G., 1993a. Determination of estuary parameters on the basis of Lagrangian analysis. Journal of Hydraulic Engineering 119(5), 628-643.

Savenije, H.H.G., 1993b. Composition and driving mechanisms of longitudinal tidal average salinity dispersion in estuaries. Journal of Hydrology 144, 127-141.

Savenije, H.H.G., 1993c. Predictive model for salt intrusion in estuaries. Journal of Hydrology 148, 203-218.

Savenije, H.H.G., 1998. Analytical expression for tidal damping in alluvial estuaries. Journal of Hydraulic Engineering 124(6), 615-618.

Savenije, H.H.G., 2001. A simple analytical expression to describe tidal damping or amplification. Journal of Hydrology 243, 205-215.

Savenije, H.H.G., 2003. The width of a bankfull channel; Lacey's formula explained. Journal of Hydrology 276(1-4), 176-183.

Savenije, H.H.G., and Veling, E.J.M., 2005. Relation between tidal damping and wave celerity in estuaries. Journal of Geophysical Research 110, C04007, 1-10.

Savenije, H.H.G., 2005. Salinity and tides in alluvial estuaries. Elsevier, Amsterdam, 197 pp.

Savenije, H.H.G., 2006. Comment on "A note on salt intrusion in funnel-shaped estuaries: Application to the Incomati estuary, Mozambique" by Brockway et al. 2006. Estuarine, Coastal and Shelf Science 68, 703-706.

Savenije, H.H.G., Haas, J., Toffolon, Marco and Veling, E.J.M., 2007. Analytical description of tidal dynamics in convergent estuaries. Submitted to Journal of Geophysical Research.

Schijf, J.B., and Schönfeld, J.C., 1953. Theoretical considerations on the motion of salt and fresh water. Proceedings of Minnesota International Hydraulics Convention, Minneapolis, Minnesota, pp. 321-333.

Signell, R.P., and Geyer, W.R., 1990. Numerical simulation of tidal dispersion around a coastal headland. In R.T. Cheng (ed.), Residual currents and long-term transport in estuaries and bays. Lecture notes on Coastal and Estuarine studies. No. 38. Springer-Verlag, New York, pp. 210-222.

Simmons, H.B., 1955. Some effects of upland discharge on estuarine hydraulics. Proceedings Amer. Soc. Civil Eng. (ASCE) 81, paper no. 792, 14 pp.

Simpson, J.H., Brown, J., Matthews, J., and Allen, G., 1990. Tidal straining, density currents, and stirring in the control of estuarine stratification. Estuaries 13, 125-132.

Speer, P.E., and Aubrey, D.G., 1985. A study of non-linear tidal propagation in shallow inlet/estuarine systems: Part II-theory. Estuarine, Coastal and Shelf Science 21, 207-224.

Stelling, G.S., 1984. On the construction of computational methods for shallow water flow problems. Rijkswaterstaat Communications, No. 35, Rijkswaterstaat, The Hague, The Netherlands.

Stelling, G.S. and Leendertse, J.J., 1991. Approximation of convective processes by cyclic ACI methods. Proceedings 2nd ASCE Conference on Estuarine and Coastal Modelling, Tampa, USA, pp. 771-782.

Stigter, C., and Siemons J., 1967. Calculation of longitudinal salt distribution in estuaries as function of time. Delft Hydraulics Laboratory Publication 52, Delft, The Netherlands.

de Swart, H.E., De Jonge, V.N., and Vosbeek, M., 1997. Application of the tidal random walk model to calculate water dispersion coefficients in the Ems estuary. Estuarine, Coastal and Shelf Science 45, 123–133.

Sylaios, G., and Boxall, S. R., 1998. Residual currents and flux estimates in a partially mixed estuary. Estuarine, Coastal and Shelf Science 46, 671–682.

Tang, D.T., 2002. Study on problem of multi-sources water system (Study site: The Mekong Delta and the Eastern-south area of Vietnam). Dr. Eng. Thesis, Southern Institute for Water Resources Research, Ho Chi Minh City (in Vietnamese).

Tee, K.T., 1978. Tide-induced residual current, a 2-D non-linear numerical tidal model. Journal of Marine Research 34, 603-628.

Thatcher, M.L., and Harleman, D.R.F., 1972. A mathematical model for the prediction of unsteady salinity intrusion in estuaries. R.M. Parsons Laboratory Report, No. 144, MIT, Cambridge, Massachusetts.

Thatcher, M.L., and Harleman, D.R.F., 1981. Long-term salinity calculation in Delaware estuary. Journal of the Environmental Engineering Division (ASCE) 107, 11-27.

Toffolon, M., Vignoli, G., and Tubino, M., 2006. Relevant parameters and finite amplitude effects in estuarine hydrodynamics. Journal of Geophysical Research 111, C10014. doi:10.1029/2005JC003104.

Trinh, T.L., and Nguyen, A.N., 2005. Socio-economic development and potential environment in the Mekong Delta, Vietnam. Proceedings of the International Symposium on Sustainable development in the Mekong river basin, Ho Chi Minh City, Vietnam, pp. 8-16.

Uncles, R.J., and Jordan, M.B., 1979. Residual fluxes of water and salt at two stations in the Severn estuary. Estuarine and Coastal Marine Science 9, 287–302.

Uncles, R.J., Elliott, R.C.A., and Weston, S.A., 1985. Dispersion of salt and suspended sediment in a partly mixed estuary. Estuaries 8, 256-269.

Van Dam, G.C., and Schönfeld, J.C., 1967. Experimental and theoretical work in the field of turbulent diffusion, performed with regard to the Netherlands estuaries and coastal regions of the North Sea. General Assembly IUGG, Berne, 1967/Rijkswaterstaat, The Netherlands, Report MFA 6807.

Van den Burgh, P., 1972. Ontwikkeling van een methode voor het voorspellen van zoutverdelingen in estuaria, kanalen en zeeen. Rijkwaterstaat Rapport, pp. 10-72 (in Dutch).

Van Eck, B., 1999. The Scheldt-atlas: an overview of an estuary. Rijkwaterstaat publication (in Dutch).

Van de Kreeke, J., and Zimmerman, J.T.F., 1990. Gravitational circulation in well-mixed and partially-mixed estuaries, , Ocean Engineering Science: The Sea 9(A), 495-522.

Van Maldegem D.C., Mulder H.P.J., and Langerak A., 1993. A cohesive sediment balance in the Scheldt estuary. Netherlands Journal of Aquatic Ecology 27(2-4), 247-256.

Van Os, A.G., and Abraham, G., 1990. Density currents and salt intrusion. Lecture Note for the Hydraulic Engineering course at IHE-Delft, Delft Hydraulics, Delft, The Netherlands.

Van Veen, J., 1950. Ebb and flood channel systems in the Netherlands tidal waters. Journal of the Royal Dutch Geographical Society (KNAG) 67, 303–325 (in Dutch, English summary). Republished, translated and annotated by Delft University of Technology, 2001.

Van Veen, J., van der Spek, A.J.F., Stive, M.J.F., and Zitman, T., 2005. Ebb and flood channel systems in the Netherlands tidal waters. Journal of Coastal Research 21(6), 1107-1120.

Verlaan, P.A.J., 1998. Mixing of marine and fluvial particles in the Scheldt estuary. Ph.D. thesis. Delft University of Technology, The Netherlands, 205 pp.

Vongvisessomjai, S., 1987. Interaction of tide and river flow. Journal of Waterway, Port, Coastal and Ocean Engineering 115(1), 86-104.

Wang, Z.B., and Winterwerp, J.C., 2001. On the stability of ebb-flood channel system. Proceedings 2nd IAHR Symposium on River, Coastal and Estuarine Morphodynamics, Obihiro, Japan, pp. 515-524.

Wang, Z.B., Jeuken, MC.J.L., Gerritsen, H., De Vriend, H.J., and Kornman, B.A., 2002. Morphology and asymmetry of the vertical tide in the Westerschelde estuary. Continental Shelf Research 22, 2599-2609.

Wang, Z.B., Jeuken, MC.J.L., and Kornman, B.A., 2003. A model for predicting dredging requirement in the WesterSchelde. Proceedings of International Conference on Estuaries and Coasts, Hangzhou, China, pp. 429-435.

Winterwerp, J.C., 1983. The decomposition of mass transport in narrow estuaries. Estuarine, Coastal and Shelf Science 16, 627–639.

Winterwerp, J.C., Wang, Z.B., Stive, M.J.F., Arends, A., Jeuken, C., Kuiper, C., and Thoolen, P.M.C., 2001. A new morphological schematisation of the Western Scheldt estuary, the Netherlands, Proceedings 2nd IAHR Symposium on River, Coastal and Estuarine Morphodynamics, Obihiro, Japan, pp. 525-534.

Winterwerp. J.C., Wang, Z.B., Kaaij, T., Verelst, K., Bijlsma, A., Meersschaut, Y., and Sas, M., 2006. Flow velocity profiles in the Lower Scheldt estuary. Ocean Dynamics 56 (3-4), 284-294.

Wolanski, E.J.I., and Heron, M., 1984. Island wakes in shallow coastal waters. Journal of Geophysical Research 89, 10553-10569.

Wolanski, E., Huan, N., Dao, L., Nhan, L., and Thuy, N., 1996. Fine sediment dynamics in the Mekong River estuary, Vietnam. Estuarine, Coastal and Shelf Science 43, 565-582.

Wolanski, E., Nguyen, H.N., and Spagnol, S., 1998. Fine sediment dynamics in the Mekong River estuary in the dry season. Journal of Coastal Research 14, 472-482.

Wolanski, E., and Nguyen, H.N., 2005. Oceanography of the Mekong River Estuary. In: Chen, Z., Saito, Y., Goodbred, S.L. (Eds.), Mega-Deltas of Asia-Geological Evolution and Human Impact. China Ocean Press, Beijing, pp. 113-115.

Zienkiewicz, O.C., and Taylor, R.L., 1989. The finite element method, Volume 1. New York, McGraw-Hill, 648 pp.

Zimmerman, J.T.F., 1976. Mixing and flushing of tidal embayments in the Western Dutch Wadden Sea, part II: Analysis of mixing processes. Netherlands Journal of Sea Research 10(4), 397-439.

Zimmerman, J.T.F., 1981. Dynamics, diffusion and geomorphological significance of tidal residual eddies. Nature 290, 549-555.

Zimmerman, J.T.F., 1986. The tidal whirlpool: a review of horizontal dispersion by tidal and residual currents. Netherlands Journal of Sea Research 20(2/3), 133-154.

LIST OF FIGURES

LIST OF TABLES

NOTATIONS

Symbols

a	Area convergence length (L)
A	Cross section area (L^2)
A_0	Cross sectional area at the estuary mouth (L^2)
B	Width convergence length (L)
B	Estuarine channel width (L)
B_0	Width at the estuary mouth (L)
c	Tidal wave celerity (LT^{-1})
c_i	x-dependent coefficient equal to the ratio between dispersion coefficient and the freshwater velocity (-)
c_0	Classical tidal wave celerity (LT^{-1})
C	Chezy coefficient ($L^{0.5}T^{-1}$)
D	Dispersion coefficient (L^2T^{-1}) (Chapter 3, 4 and 6)
D	Damping term (-) (Chapter 2 and 5)
D_m	Coefficient of proportionality (L^2T^{-1})
D_{TA}	Tidal and cross-sectional average dispersion (L^2T^{-1})
D_0^{HWS}	HWS dispersion coefficient at the mouth (L^2T^{-1})
D_0^{LWS}	LWS dispersion coefficient at the mouth (L^2T^{-1})
e_p	Pumping efficiency (-)
E	Tidal excursion (L)
E_D	Estuary number (-)
E_0	Tidal excursion at the mouth (L)
f	Darcy-Weisbach's roughness (-)
F	Solute mass flux vector per unit width ($ML^{-1}T^{-1}$)
F	Froude number (-) (Chapter 2)
F_1, F_2	Hydrostatic forces ($ML^{-2}T^{-2}$)
F_D	Estuarine densimetric Froude number (-)
g	Acceleration due to gravity (LT^{-2})
h_0	Depth at the estuary mouth (L)
H	Tidal range (L)

$\overline{h}$	Tidal average depth (L)
k	Bed friction (-)
k_A	Arons and Stommel constant (-)
K	Van der Burgh's coefficient (-)
l_0	Mixing length scale (L)
L	Salt intrusion length (L)
L_i	Salt intrusion length at HWS, LWS or TA (L)
m	Width variation coefficient (-)
M	Momentum driven by gravitational circulation ($ML^{-1}T^{-2}$)
n	Depth variation coefficient (-)
N	Volumetric ratio (-)
N_R	Estuarine Richardson number (-)
P_t	Tidal prism (L^3)
Q	Discharge (L^3T^{-1})
Q	Instantaneous rate of water transport (L^3T^{-1}) (Chapter 2)
Q_f	Freshwater discharge (L^3T^{-1})
Q_S	Instantaneous rate of salt transport (L^3T^{-1})
Q_t	Tidal discharge (L^3T^{-1})
$\langle Q \rangle$	Tidal average (residual) water transport (L^3T^{-1})
$\langle Q_S \rangle$	Tidal average (residual) salt transport (L^3T^{-1})
r_s	Storage width ratio (-)
R_S	Source term (L^2T^{-1})
s	Mass concentration of diffusing solute (ML^{-3})
S	Steady-state salinity (-)
S_f	Freshwater salinity (-)
S_{i0}	Salinity at the river mouth (x=0) for the HWS, LWS or TA conditions (-)
t	Time (T)
T	Tidal period (T)
T_s	System response time (T)
u	Flow velocity (LT^{-1})
u_0	Freshwater velocity at the estuary mouth (LT^{-1})
u_f	Freshwater velocity (LT^{-1})
U	Mean cross-sectional flow velocity (LT^{-1})

U_0	Velocity scale (LT^{-1})
V_1	Eulerian residual current (LT^{-1})
V_2	Stokes drift current (LT^{-1})
V_L	Lagrangean residual current (LT^{-1})
$V_{S,1}$	Cross-sectional residual flux of salt transport due to the residual transport of water (LT^{-1})
$V_{S,2}$	Cross-sectional residual flux of salt transport due to tidal pumping (LT^{-1})
$V_{S,3}$	Cross-sectional residual flux of salt transport due to gravitational circulation (LT^{-1})
x	Distance from the mouth (L)
z	Vertical axis (L)
Z_b	Mean cross-sectional bottom elevation (L)
α	Tidal Froude number (-)
α_S	Shape factor (-)
δ	Tidal damping number (-)
ε	Phase lag (T)
η	Tidal amplitude (L)
λ	Celerity number (-)
μ	Velocity number (-)
υ	Tidal velocity amplitude (LT^{-1})
υ_0	Tidal velocity amplitude at the river mouth (LT^{-1})
$\Delta\rho$	Density difference between seawater and river water (MT^{-3})
ρ	Density of fresh water (MT^{-3})
γ	Estuary shape number (-)
χ	Friction number (-)
ω	Angular velocity (T^{-1})

Abbreviations

LWS	Low water slack
HWS	High water slack
LW	Low water
HW	High water
TA	Tidal average

ACKNOWLEDGEMENTS

It has been four years since I started this Ph.D. study in February 2004. Looking back on the study period, I always remind myself that the study would certainly have become a "mission impossible" without direct and indirect help, support and encouragement of many people. Therefore, I would like to take this opportunity to express my heartfelt thanks and acknowledgements to them.

First of all, I would like to express my special thanks to my promotor and supervisor, Prof. Hubert H.G. Savenije. Huub, without your support, your enthusiasm, your countless help and encouragement, this study would have lasted forever. I greatly appreciate your time devoted to help me during the study. I treasured the moment we were together in Perth in 2005 when I shifted to a higher gear in my research after a slow beginning. I really enjoyed the dinners at your house, together, of course, with your lovely wife Heleen. I also enjoyed our fieldtrips to the Mekong Delta in 2005 and especially in 2006 with the group of Dutch students.

Secondly, I thank Ir. Mick van der Wegen for being my mentor and my close friend during the last four years. Mick, I am grateful for the essential discussions on hydrodynamic modeling and code programming with you, as well as many talks besides research interests. I would like to thank Dr. Zheng Bing Wang for a number of interesting discussions during my study, for giving permission to use the Scheldt estuary schematization in Sobek-Re and for his comments on my papers and draft thesis. Prof. Dano Roelvink is warmly thanked for fruitful discussions in hydrodynamic modeling and for giving permission to use the Scheldt estuary schematization in Delft3D.

I would like to express my gratitude to the committee members, Prof. Marcel Stive, Prof. Guus Stelling, Prof. Nick van der Giesen for agreeing to be in the committee, for reading the draft thesis and giving their constructive comments.

My thanks go to Ir. Adri Verwey and Ir. Theo van der Kijn (WL|Delft Hydraulics) for several interesting talks on hydrodynamic modeling. Jennifer Haas whom I co-supervised during her M.Sc. study is thanked for contributing parts of chapter 5.

I should acknowledge Dr. Tang Duc Tang, Southern Institute for Water Resources Research in Vietnam (SIWRR), for his kind support during my study and his permission to use the Mekong Delta schematization in Mike11 and the delta's hydrological database. I also learned much from him about the hydrodynamic characteristics of the Mekong Delta. I would also like to express my sincere gratitude to Prof. Le Sam and Mr. Nguyen Van Sang, SIWRR, for their permission to use the salinity database of the Mekong Delta. Pham Duc Nghia is thanked for helping me in several modeling aspects of the Mekong river system.

This study would have not been carried out without sufficient financial support. I would like to deeply acknowledge the Netherlands Fellowship Programmes (NFP), the Vietnamese Ministry of Education and Training, and the Section of Water Resources (CiTG, TU Delft) for funding my research.

My colleagues at UNESCO-IHE Institute for Water Education and Section of Water Resources are thanked for their friendly working environment. Nico, Miriam, Zhang, Fabrizio, Sebastian, Emmanuel, Kanapoj, Marloes, Hessel, Robert, Susan, thank you for the good atmosphere, many nice talks and jokes about other aspects beside research interests. Jolanda and Hanneke, thank you for helping me on financial and practical aspects of the Ph.D. study.

Many thanks are given to Vietnamese friends in Delft for their pleasant social environment, which lightened my homesickness, especially for the ping-pong group every Friday afternoon giving me essential physical practice to reduce working pressure. I would like to thank people in the Vietnamese Embassy in the Netherlands for their support during my six-and-half-year period in the Netherlands.

Finally, I am absolutely indebted to my parents for their love, encouragement and their patience during my study period in the Netherlands. I am glad to always feel the great love and support of my parents and my sister's family even across a 10,000 km distance.

Nguyen Anh Duc

Delft, February 2008.

ABOUT THE AUTHOR

Nguyen Anh Duc was born on 4 July 1976 in Hanoi, Vietnam. He stayed in Hanoi until the first year of his secondary education. In 1992, he moved to Ho Chi Minh City with his family. He graduated there from the Gia Dinh secondary school in 1994. He started his academic education at the Ho Chi Minh City University of Architecture, majoring in Building Construction and Industrial Structures. He graduated in 1999 with a thesis entitled "Structural computations for a multi-story building". Shortly after, he was employed as a junior researcher by the Center for Hydro-informatics, Southern Institute for Water Resources Research in Ho Chi Minh City. Between 1999 and 2000, he was responsible for designing and supervising the construction of sluices and dikes in the Western Sea water resources project in the Mekong Delta, Vietnam.

In 2001, he obtained a post-graduate scholarship to study at the UNESCO-IHE Institute for Water Education in the Netherlands, which was covered partially by the Vietnamese Government and partially by the Netherlands Fellowship Programmes (NFP). He participated in the Hydraulic Engineering programme and obtained his M.Sc. degree (with distinction) in June 2003, on a thesis entitled "Flood management strategies: A case study in the lower Dong Nai – Sai Gon river system, Vietnam". After graduation, he remained at the Department of Management and Institutions (UNESCO-IHE) as a research assistant until the beginning of 2004.

In February 2004, he started his Ph.D. research at UNESCO-IHE on salt intrusion, mixing and tides in alluvial estuaries, which resulted in the present dissertation. The research was funded by NFP, the Vietnamese Government and the Section of Water Resources (Delft University of Technology). In the course of the Ph.D. research, in 2005, he had a 2-month visit as a guest researcher to the Center of Water Research, University of Western Australia. He published three papers in international journals and presented several papers at national and international conferences in The Hague (The Netherlands), Vienna (Austria), Varna (Bulgaria), Alexandria (Egypt) and Phoenix (USA). In addition, he co-supervised two M.Sc. studies on salinity intrusion and tides.

For Product Safety Concerns and Information please contact our EU
representative GPSR@taylorandfrancis.com
Taylor & Francis Verlag GmbH, Kaufingerstraße 24, 80331 München, Germany

www.ingramcontent.com/pod-product-compliance
Ingram Content Group UK Ltd.
Pitfield, Milton Keynes, MK11 3LW, UK
UKHW051836180425
457613UK00023B/1281

* 9 7 8 0 4 1 5 4 7 1 2 2 0 *